IET POWER AND ENERGY SERIES 56

Condition Monitoring of Rotating Electrical Machines

Other volumes in this series:

Volume 1   **Power circuit breaker theory and design** C.H. Flurscheim (Editor)
Volume 4   **Industrial microwave heating** A.C. Metaxas and R.J. Meredith
Volume 7   **Insulators for high voltages** J.S.T. Looms
Volume 8   **Variable frequency AC motor drive systems** D. Finney
Volume 10   **SF6 switchgear** H.M. Ryan and G.R. Jones
Volume 11   **Conduction and induction heating** E.J. Davies
Volume 13   **Statistical techniques for high voltage engineering** W. Hauschild and W. Mosch
Volume 14   **Uninterruptible power supplies** J. Platts and J.D. St Aubyn (Editors)
Volume 15   **Digital protection for power systems** A.T. Johns and S.K. Salman
Volume 16   **Electricity economics and planning** T.W. Berrie
Volume 18   **Vacuum switchgear** A. Greenwood
Volume 19   **Electrical safety: a guide to causes and prevention of hazards** J. Maxwell Adams
Volume 21   **Electricity distribution network design**, 2nd edition, E. Lakervi and E.J. Holmes
Volume 22   **Artificial intelligence techniques in power systems** K. Warwick, A.O. Ekwue and R. Aggarwal (Editors)
Volume 24   **Power system commissioning and maintenance practice** K. Harker
Volume 25   **Engineers' handbook of industrial microwave heating** R.J. Meredith
Volume 26   **Small electric motors** H. Moczala *et al.*
Volume 27   **AC–DC power system analysis** J. Arrillaga and B.C. Smith
Volume 29   **High voltage direct current transmission**, 2nd edition J. Arrillaga
Volume 30   **Flexible AC Transmission Systems (FACTS)** Y-H. Song (Editor)
Volume 31   **Embedded generation** N. Jenkins *et al.*
Volume 32   **High voltage engineering and testing**, 2nd edition H.M. Ryan (Editor)
Volume 33   **Overvoltage protection of low-voltage systems**, revised edition P. Hasse
Volume 34   **The lightning flash** V. Cooray
Volume 35   **Control techniques drives and controls handbook** W. Drury (Editor)
Volume 36   **Voltage quality in electrical power systems** J. Schlabbach *et al.*
Volume 37   **Electrical steels for rotating machines** P. Beckley
Volume 38   **The electric car: development and future of battery, hybrid and fuel-cell cars** M. Westbrook
Volume 39   **Power systems electromagnetic transients simulation** J. Arrillaga and N. Watson
Volume 40   **Advances in high voltage engineering** M. Haddad and D. Warne
Volume 41   **Electrical operation of electrostatic precipitators** K. Parker
Volume 43   **Thermal power plant simulation and control** D. Flynn
Volume 44   **Economic evaluation of projects in the electricity supply industry** H. Khatib
Volume 45   **Propulsion systems for hybrid vehicles** J. Miller
Volume 46   **Distribution switchgear** S. Stewart
Volume 47   **Protection of electricity distribution networks**, 2nd edition J. Gers and E. Holmes
Volume 48   **Wood pole overhead lines** B. Wareing
Volume 49   **Electric fuses**, 3rd edition A. Wright and G. Newbery
Volume 50   **Wind power integration: connection and system operational aspects** B. Fox *et al.*
Volume 51   **Short circuit currents** J. Schlabbach
Volume 52   **Nuclear power** J. Wood
Volume 53   **Condition assessment of high voltage insulation in power system equipment** R.E. James and Q. Su
Volume 905 **Power system protection**, 4 volumes

Condition Monitoring of Rotating Electrical Machines

Peter Tavner, Li Ran, Jim Penman and Howard Sedding

The Institution of Engineering and Technology

Published by The Institution of Engineering and Technology, London, United Kingdom

© 2008 The Institution of Engineering and Technology

First published 2008

The Institution of Engineering and Technology
Michael Faraday House
Six Hills Way, Stevenage
Herts, SG1 2AY, United Kingdom

www.theiet.org

While the author and the publishers believe that the information and guidance given in this work are correct, all parties must rely upon their own skill and judgement when making use of them. Neither the author nor the publishers assume any liability to anyone for any loss or damage caused by any error or omission in the work, whether such error or omission is the result of negligence or any other cause. Any and all such liability is disclaimed.

British Library Cataloguing in Publication Data
A catalogue record for this product is available from the British Library

ISBN 978-0-86341-739-9

Typeset in India by Newgen Imaging Systems (P) Ltd, Chennai
Printed in the UK by Athenaeum Press Ltd, Gateshead, Tyne & Wear

This book is dedicated to Ying Lu

Considerate la vostra semenza;
Fatti non foste a viver come bruti,
Ma per seguir virtute e conoscenza.
Consider the seed from which you were made;
You were not made to live like brutes,
But to pursue virtue and knowledge.

Dante's Inferno,
Canto XXVI, lines 118–120, Ulysses

The photograph is of 3 phase, 290 kW, 6.6 kV, 60 Hz, 6-pole. 1188 rev/min squirrel cage induction motors, manufactured by ATB Laurence Scott Ltd at Norwich in the UK, driving pumps on the offshore Buzzard platform in the North Sea.

Contents

Preface   xiii

Acknowledgments   xvii

Nomenclature   xix

1   Introduction to condition monitoring   1
  1.1   Introduction   1
  1.2   The need for monitoring   4
  1.3   What and when to monitor   7
  1.4   Scope of the text   9
  1.5   References   10

2   Construction, operation and failure modes of electrical machines   13
  2.1   Introduction   13
  2.2   Materials and temperature   14
  2.3   Construction of electrical machines   16
     2.3.1   General   16
     2.3.2   Stator core and frame   18
     2.3.3   Rotors   18
     2.3.4   Windings   18
     2.3.5   Enclosures   20
     2.3.6   Connections   26
     2.3.7   Summary   26
  2.4   Structure of electrical machines and their types   26
  2.5   Machine specification and failure modes   33
  2.6   Insulation ageing mechanisms   35
     2.6.1   General   35
     2.6.2   Thermal ageing   36
     2.6.3   Electrical ageing   36
     2.6.4   Mechanical ageing   37
     2.6.5   Environmental ageing   38
     2.6.6   Synergism between ageing stresses   39
  2.7   Insulation failure modes   39
     2.7.1   General   39
     2.7.2   Stator winding insulation   40
     2.7.3   Stator winding faults   45

	2.7.4	Rotor winding faults	50
2.8		Other failure modes	54
	2.8.1	Stator core faults	54
	2.8.2	Connection faults (high-voltage motors and generators)	54
	2.8.3	Water coolant faults (all machines)	56
	2.8.4	Bearing faults	56
	2.8.5	Shaft voltages	56
2.9		Conclusion	59
2.10		References	59

3   Reliability of machines and typical failure rates   **61**
3.1	Introduction	61
3.2	Definition of terms	61
3.3	Failure sequence and effect on monitoring	63
3.4	Typical root causes and failure modes	65
	3.4.1 General	65
	3.4.2 Root causes	65
	3.4.3 Failure modes	66
3.5	Reliability analysis	66
3.6	Machinery structure	69
3.7	Typical failure rates and MTBFs	71
3.8	Conclusion	75
3.9	References	76

4   Instrumentation requirements   **79**
4.1	Introduction	79
4.2	Temperature measurement	81
4.3	Vibration measurement	88
	4.3.1 General	88
	4.3.2 Displacement transducers	89
	4.3.3 Velocity transducers	91
	4.3.4 Accelerometers	92
4.4	Force and torque measurement	94
4.5	Electrical and magnetic measurement	97
4.6	Wear and debris measurement	100
4.7	Signal conditioning	102
4.8	Data acquisition	104
4.9	Conclusion	106
4.10	References	106

5   Signal processing requirements   **109**
5.1	Introduction	109
5.2	Spectral analysis	110
5.3	High-order spectral analysis	115

5.4	Correlation analysis	116
5.5	Signal processing for vibration	118
	5.5.1 General	118
	5.5.2 Cepstrum analysis	118
	5.5.3 Time averaging and trend analysis	120
5.6	Wavelet analysis	121
5.7	Conclusion	125
5.8	References	125

6   Temperature monitoring   **127**
6.1	Introduction	127
6.2	Local temperature measurement	127
6.3	Hot-spot measurement and thermal images	132
6.4	Bulk measurement	132
6.5	Conclusion	134
6.6	References	134

7   Chemical monitoring   **137**
7.1	Introduction	137
7.2	Insulation degradation	137
7.3	Factors that affect detection	138
7.4	Insulation degradation detection	142
	7.4.1 Particulate detection: core monitors	142
	7.4.2 Particulate detection: chemical analysis	146
	7.4.3 Gas analysis off-line	148
	7.4.4 Gas analysis on-line	149
7.5	Lubrication oil and bearing degradation	152
7.6	Oil degradation detection	153
7.7	Wear debris detection	153
	7.7.1 General	153
	7.7.2 Ferromagnetic techniques	154
	7.7.3 Other wear debris detection techniques	155
7.8	Conclusion	157
7.9	References	157

8   Vibration monitoring   **159**
8.1	Introduction	159
8.2	Stator core response	159
	8.2.1 General	159
	8.2.2 Calculation of natural modes	161
	8.2.3 Stator electromagnetic force wave	164
8.3	Stator end-winding response	167
8.4	Rotor response	168
	8.4.1 Transverse response	168
	8.4.2 Torsional response	171

8.5	Bearing response		173
	8.5.1	General	173
	8.5.2	Rolling element bearings	173
	8.5.3	Sleeve bearings	175
8.6	Monitoring techniques		176
	8.6.1	Overall level monitoring	177
	8.6.2	Frequency spectrum monitoring	179
	8.6.3	Faults detectable from the stator force wave	182
	8.6.4	Torsional oscillation monitoring	183
	8.6.5	Shock pulse monitoring	187
8.7	Conclusion		189
8.8	References		189
9	**Electrical techniques: current, flux and power monitoring**		**193**
9.1	Introduction		193
9.2	Generator and motor stator faults		193
	9.2.1	Generator stator winding fault detection	193
	9.2.2	Stator current monitoring for stator faults	193
	9.2.3	Brushgear fault detection	194
	9.2.4	Rotor-mounted search coils	194
9.3	Generator rotor faults		194
	9.3.1	General	194
	9.3.2	Earth leakage faults on-line	195
	9.3.3	Turn-to-turn faults on-line	196
	9.3.4	Turn-to-turn and earth leakage faults off-line	204
9.4	Motor rotor faults		207
	9.4.1	General	207
	9.4.2	Airgap search coils	207
	9.4.3	Stator current monitoring for rotor faults	207
	9.4.4	Rotor current monitoring	210
9.5	Generator and motor comprehensive methods		212
	9.5.1	General	212
	9.5.2	Shaft flux	213
	9.5.3	Stator current	217
	9.5.4	Power	217
	9.5.5	Shaft voltage or current	219
	9.5.6	Mechanical and electrical interaction	221
9.6	Effects of variable speed operation		221
9.7	Conclusion		224
9.8	References		224
10	**Electrical techniques: discharge monitoring**		**229**
10.1	Introduction		229
10.2	Background to discharge detection		229
10.3	Early discharge detection methods		231

		10.3.1	RF coupling method	231
		10.3.2	Earth loop transient method	233
		10.3.3	Capacitive coupling method	235
		10.3.4	Wideband RF method	236
		10.3.5	Insulation remanent life	236
	10.4	Detection problems		238
	10.5	Modern discharge detection methods		239
	10.6	Conclusion		241
	10.7	References		241

11 Application of artificial intelligence techniques — **245**
	11.1	Introduction		245
	11.2	Expert systems		246
	11.3	Fuzzy logic		250
	11.4	Artificial neural networks		253
		11.4.1	General	253
		11.4.2	Supervised learning	254
		11.4.3	Unsupervised learning	256
	11.5	Conclusion		260
	11.6	References		261

12 Condition-based maintenance and asset management — **263**
	12.1	Introduction		263
	12.2	Condition-based maintenance		263
	12.3	Life-cycle costing		265
	12.4	Asset management		265
	12.5	Conclusion		267
	12.6	References		268

Appendix Failure modes and root causes in rotating electrical machines — **269**

Index — **277**

Preface

Condition monitoring of engineering plant has increased in importance as more engineering processes become automated and the manpower needed to operate and supervise plant is reduced. However, electrical machinery has traditionally been thought of as reliable and requiring little attention, except at infrequent intervals when the plant is shut down for inspection. Indeed the traditional application of fast-acting protective relays to electrical machines has rather reduced the attention that operators pay to the equipment.

Rotating electrical machines, however, are at the heart of most engineering processes and as they are designed to tighter margins there is a growing need, for reliability's sake, to monitor their behaviour and performance on-line. This book is a guide to the techniques available. The subject of condition monitoring of electrical machines as a whole covers a very wide field including rotating machines and transformers. To restrict the field the authors deal with rotating machines only and with techniques that can be applied when those machines are in operation, neglecting the many off-line inspection techniques.

The first edition of this book, *Condition Monitoring of Electrical Machines*, was written by Peter Tavner and Jim Penman and published in 1987 by Research Studies Press, with the intention of bringing together a number of strands of work active at that time from both industry and academia. In academia there was a growing confidence in the mathematical analysis of electrical machines, in computer modelling of complex equivalent circuits and in the application of finite-element methods to predict their magnetic fields. In industry there was growing interest in providing better monitoring for larger electrical machines as rising maintenance costs competed with the heavy financial impact of large machine failures.

The original book was primarily aimed at larger machines involved in energy production, such as turbine generators and hydro generators, boiler feed pumps, gas compressors and reactor gas circulators. This was because at that time those were the only plant items costly enough to warrant condition-monitoring attention. It also reflected the fact that one of the authors worked in the nationalised generating utility, colouring his approach to the subject.

The original book showed that, in respect of condition monitoring, electrical machines are unusual when compared with most other energy conversion rotating plant. The all-embracing nature of the electromagnetic field in the energy conversion process, which is the *raison d'être* of the electrical machine, enables operators to infer far more about their operation from their terminal conditions than is usually the case with non-electrical rotating machinery. In this earlier work the authors were inspired by a much earlier book by Professor Miles Walker, *The Diagnosing of Trouble in Electrical Machines*, first published in 1921.

Our book covered the elemental aspects of electrical machine condition monitoring but exposed a number of important facets of understanding that have subsequently lead to a great deal of further work, namely

- the electromagnetic behaviour of electrical machines,
- the dynamic behaviour of electrical machines, particularly associated with the control now available with modern power electronics,
- the behaviour of electrical machine insulation systems.

Each of these facets have now matured and are a rich source of fundamental knowledge that has been related to the behaviour of machines in their operating state, especially under fault conditions. Two examples of this further work are Professor Peter Vas', *Parameter Estimation, Condition Monitoring and Diagnosis of Electrical Machines*, published in 1996; and Greg C. Stone's, *Electrical Insulation for Rotating Machines, Design, Evaluation, Ageing, Testing and Repair*, published in 2004.

The economics of industry has also changed, particularly as result of the privatisation and deregulation of the energy industry in many countries, placing far greater emphasis on the importance of reliable operation of plant and machinery, throughout the whole life cycle, regardless of its first capital cost.

Finally the availability of advanced electronics and software in powerful instrumentation, computers and digital signal processors has simplified and extended our ability to instrument and analyse machinery, not least in the important area of visualising the results of complex condition-monitoring analysis. As a result, condition monitoring is now being applied to a wider range of systems, from fault-tolerant drives of a few hundred watts in the aerospace industry, to machinery of several hundred megawatts in major capital plant. The value of the fundamental contribution to these advances by many analysts over the last 20 years cannot be underestimated and they will play a major part in the future.

In this new book, *Condition Monitoring of Rotating Electrical Machines*, the original authors have been joined by their colleague Dr Li Ran, an expert in power electronics and control, and Dr Howard Sedding, an expert in the monitoring of electrical insulation systems. Together we have decided to build upon the earlier book, retaining the same limits we set out at the start of this preface, merging our own experience with that of the important machine analysts through the years to bring the reader a thoroughly up-to-date but practicable set of techniques that reflect the work of the last 20 years. The book is aimed at professional engineers in the energy, process engineering and manufacturing industries, research workers and students. We have placed an additional limit on the book and that is to consider the machine itself rather than its control systems. While recognising the enormous growth of the application of electronic variable speed drives in industry, we do not deal with their specific problems except in passing. We acknowledge that this is important for future growth but leave this area of investigation to a future author.

The examples of faults have concentrated on conventional machines rather than the emerging brushless, reluctance, permanent magnet and unusual topology machines. This is because the industry is still dominated by these conventional machines. The 'failure mode' information for newer designs has not yet emerged but will be based on earlier machine experience. In this edition we have omitted case studies because the range of application of condition-monitoring techniques on electrical machines is now so wide and complex that it is difficult to select appropriate applications from which general conclusions can be drawn.

We have introduced a 'Nomenclature' section and extended the references to cover major recent journal papers and books that have illuminated the subject, including some of the older seminal works, which still deserve scrutiny. The authors have also taken the opportunity to correct errors in the previous book, rearrange the material presented and add important information about failure mechanisms, reliability, instrumentation, signal processing and the management of rotating machine assets as these factors critically affect the way in which condition monitoring needs to be applied. Finally, the diagrams and photographs representing the machines, the monitoring systems and the signal processing used have been updated where appropriate.

Peter Tavner
Durham University, 2006

Acknowledgments

Peter Tavner and Li Ran acknowledge the assistance they have had from Durham University in preparing this book, particularly from Barbara Gilderoy and Denise Norman for assisting in transferring the material from our previous edition to this new one and to Chris Orton and Julie Morgan-Dodds for carefully drawing many of the diagrams. They also acknowledge the help of students and research assistants, including Xiang Jianping, Michael Wilkinson, Fabio Spinato and Mark Knowles, for their contribution to the book through proofreading, discussions and their understanding of the problems of machine reliability and monitoring. Peter Tavner acknowledges the help of Dr Jim Bumby at Durham University for providing advice on the frequencies of vibration, current, flux and power associated with faults.

The authors acknowledge the assistance of companies who have contributed photographs and diagrams, in particular Brush Turbogenerators (Loughborough, UK; Plzeň, Czech Republic; and Ridderkirk, Netherlands), Marelli Motori S.p.A (Arzignano, Italy), Dong Feng Electrical Machinery Ltd (Deyang, China) and Laurence, Scott & Electromotors Ltd (Norwich, UK).

The photograph on the front cover is of 3 290 kW, 6.6 kV, 60 Hz, six-pole, 1 188 rev/min induction motors, manufactured by Laurence, Scott & Electromotors Ltd at Norwich in the UK, driving pumps on the offshore Buzzard platform in the North Sea.

Nomenclature

Symbol	Explanation
A	effective cross-sectional area of a coil, m^2
A	availability, $A = \text{MTBF}/(\text{MTBF} + \text{MTTR})$
$A(t)$	availability function of a population of components as a function of time
a	scaling factor of time in a mother wavelet transform
α	scale parameter in a power law expression
α_n and b_n	strain coefficients for the strain energy of the stator core
α_t	cross-sectional area of a tooth, m^2
α	scale parameter in a Weibull function
α_r	resistance temperature coefficient, degree/ohm
α_s	skew angle of a stator, degree
$B(f_1, f_2)$	bispectrum
B	radial flux density in an airgap, Tesla
b_1 or b_2	stator or rotor side instantaneous radial flux density, Tesla
b	time-shifting parameter in a mother wavelet transform
β	shape parameter in a Weibull function
β	half-angle subtended by a shorted turn, degrees
C	volumetric concentration of a degradation product in a machine
C	Carter factor to account for airgap slotting
$C(y, v)$	inverse wavelet transform
$C(t)$	cepstrum function
c	damping constant of a support system, N/m/s
D	damping factor for rotor vibrations
d	rolling element diameter, m
E	Young's modulus of a material
E_e	stored energy in an electrical system, Joules
e	specific unbalance $e = mr/M$, m
$e(t)$	instantaneous induced EMF, V
ϵ	strain in a material
θ	MTBF of a component, $\theta = 1/\lambda$ hours
θ_1	space position in the stator field, degrees
θ_2	space position in the rotor field, degrees
F	parameter from shock pulse measurement of a rolling element bearing

$F_m(\theta, t)$	forcing function on a rotor or stator expressed in circumferential angle θ and time t, N
$F(t)$	failure mode probability density, a Weibull function
F or F^{-1}	forward or backward Fourier transform
f_0	first critical or natural frequency of a rotor system, Hz
f_1 or $f_2(t)$	stator or rotor side instantaneous magnetomotive force (MMF), $N_1 I_1$ or $N_2 I_2$, ampere-turns
f_{se}	electrical supply side frequency $= 1/T$, Hz
f_{sw}	PWM switching frequency, Hz
f_m	higher m^{th} natural frequencies of the stator core, Hz
f_{sm}	mechanical vibration frequency on the stator side, Hz
f_n	the nth component of an unbalanced forcing function
f_{rm}	mechanical rotational frequency $= N/60$, Hz
G	strain gauge factor
G	degree of residual unbalance as denoted by the quantity $G = e\omega$
$G(f_k)$	generalised power spectral function of frequency, f_k, the kth harmonic
$G^*(f_k)$	complex conjugate of $G(f_k)$
$G(m)$	stiffness function of an mth natural frequencies of the stator core
$G(t_n)$	generalised periodic function of time, t_n
g	acceleration due to gravity, m/s^2
g	airgap length, mm
$g_n(z)$	natural frequency function an nth solution of the balance equation
h	heat transfer coefficient from an insulation surface, W/m^{2}K
h_t	tooth depth, m
I_1 or I_2	stator or rotor side rms current, A
i_1 or $i_2(t)$	stator or rotor side instantaneous current, A
J	polar moment of inertia of the core cylinder, joules^2
k	integer constant, indicates the stator MMF space harmonics, 1, 3, 5, 7…
k	heat transfer coefficient through an insulating material, W/mK
k	stiffness constant of a support system, N/m
k_r	reflection coefficient in the recurrent surge oscillography (RSO) test
k_c	integer number of commutator segments in a DC machine
k/nq	Hall effect constant of an electronic material
k_e	integer constant, indicates eccentricity order number, which is zero for static eccentricity and a low integer value for dynamic eccentricity
k_c	integer constant, indicates the circumferential modes in a vibrating stator core
k_l	integer constant, indicates the lengthwise modes in a vibrating stator core
k_{wn}	stator winding factor for the nth harmonic
L	active length of a core, m

L	inductance of a coil, H
ℓ	integer number of stator time harmonics or rotor winding fault harmonics
Λ	magnetic permeance
λ	instantaneous failure rate or hazard function of a component or machine, failures/component/year
$\lambda(t)$	failure rate of a component or machine varying with time, failures/component/year
M	mass of a rotating system, kg
M_s	mass of a support system, kg
m	integer constant
m	equivalent unbalance mass on a shaft, kg
μ	permeability in a magnetic field
N_1 or N_2	integer number of stator or rotor side turns of a coil
N	speed of a machine rotor, rev/min
N_r	integer number of rotor slots
N_s	integer number of stator slots
n	number of charge carriers per unit volume in a semiconductor
n	integer constant, 1, 2, 3, 4…
n_b	integer number of rolling elements in a rolling element bearing
P_1	stator side power, watt
$p_1(t)$	instantaneous stator side power, watt
p	integer number of pole pairs
Q	heat flow, watt/m^2
Q_m	maximum partial discharge recorded in partial discharge tests using a calibrated coupler, mv
q	electronic charge, coulomb
q	integer phase number
ΔR	change in resistance
R	resistance, ohms
R	shock pulse meter reading
$R(t)$	reliability or survivor function of a population of components as a function of time, failures/machine/year
$R_{ff}(t)$	auto-correlation function on a time function $f(t)$ with a delay of g t
$R_{fh}(t)$	cross-correlation function between time functions $f(t)$ and $h(t)$ with a delay of gt
R_0	resistance of a device made of the metal at 0 °C, ohm
R_T	resistance, ohm
r	effective radius of an equivalent unbalanced mass, m
r_{mean}	mean radius of a core, m
r_{airgap}	radius of airgap, m
S	constant related to the stiffness of a winding, insulation and tooth components

s	slip of an induction machine, between 0 and 1
T	torque, Nm
T	temperature, °C
T	period of a wave, sec
T_ℓ	volumetric vibration kinetic energy, joules/m^3
$T(f_1, f_2, f_3)$	trispectrum
τ_0	radial thickness of a stator core annulus, m
ρ	density of a material, kg/m^3
σ	electrical conductivity of a region, ohm.m
σ_r and σ_q	radial and tangential Maxwell stress in the airgap, N/m^2
τ_w	time duration of an overheating incident, s
τ_r	residence time of an overheating product in a machine, or leakage factor, s
τ	time delay in a correlation function, s
u	lateral displacement of a machine rotor, μm
u_r and u_θ	radial and peripheral displacements in a strained stator core, μm
V	rms voltage, volts
V	machine volume, m^3
V_ℓ	volumetric strain potential energy, joules/m^3
v	velocity of the rotor, relative to the travelling flux wave produced by the stator, m/s
$\dot{v}$	volumetric rate of production of a detectable substance, m^3/s
$\dot{v}_b$	background rate of production of the substance, m^3/s
ν	Poisson's ratio of a material
ϕ	flux, Webers
ϕ	contact angle with races of a rolling element bearing, degree
ϕ	electrical phase angle of a stator MMF wave F_s, degree
ψ	angular frequency of an electrical supply, rad/s
ψ_0	first critical or natural angular frequency of a rotor system, rad/s
ψ_{se}	electrical supply side angular frequency, rad/s
ψ_{sm}	mechanical angular vibration frequency on the stator side, rad/s
ψ_{rm}	mechanical rotational angular frequency $= 2pN/60$, rad/s
ψ_{ecc}	angular velocity of an eccentricity, rad/s
$\psi(t)$	mother wavelet function of time
W	work function for strain energy in stator core
$W(a, b)$	wavelet transform
w	weight per unit length per unit circumferential angle of a stator core cylinder, N/m
w_y, w_t, w_i and w_w	weights of a core yoke, teeth, insulation and windings, respectively, kg
X_{m2}	second harmonic magnetising reactance, ohm
X_{12}	second harmonic leakage reactance, ohm
z	longitudinal distance from the centre of a machine, m
Z_0	surge impedance of a winding, ohm

Chapter 1
Introduction to condition monitoring

1.1 Introduction

Rotating electrical machines permeate all areas of modern life at both the domestic and industrial level. The average modern home in the developed world contains 20–30 electric motors in the range 0–1 kW for clocks, toys, domestic appliances, air conditioning or heating systems. Modern cars use electric motors for windows, windscreen wipers, starting and now even for propulsion in hybrid vehicles. A modern S-series Mercedes-Benz car is reported to incorporate more than 120 separate electrical machines.

The majority of smaller applications of electrical machines do not require monitoring, the components are sufficiently reliable that they can outlive the life of the parent product. However, modern society depends, directly or indirectly, upon machines of greater rating and complexity in order to support an increased standard of living.

The electricity we use so freely is generated in power plants by machines whose rating can exceed 1 000 MW and which have evolved to a state of great sophistication. These power plants are supported by fossil fuel and nuclear energy industries that involve the transport of raw materials using pumps, compressors and conveyors in sophisticated engineering processes incorporating rotating electrical machines of powers ranging from 100 kW to 100 MW. These have been joined by a growing renewable energy industry using many of these and new techniques to extract energy from renewable sources often in combination with traditional sources.

The steel used in cars will have been rolled using large electrical machines and at an earlier stage the furnaces will have been charged using more electrical machines. Our water and waste systems are also driven by electrical machines, as are the processes that produce the raw materials for the agricultural, chemical and pharmaceutical industries. Without all these our society, as it exists at the moment, would cease to function.

The overall picture is that electrical machines come in many sizes and fulfil their function either independently or as part of a highly complex process in which all elements must function smoothly so that production can be maintained. It is the usage of electrical machinery in the latter role that has risen dramatically towards the end of the twentieth century, and there is no reason to suspect that this trend will do anything other than accelerate in the twenty-first century. However, historically the function of an individual electrical machine was seen as separable from the rest of the electromechanical system. It must be remembered that the power-to-weight ratio of electrical machines has been much lower than steam, diesel and gas engines and consequently their reliability has been much higher.

It is against this background that the basic principles of protective relaying evolved. Protection is designed to intercept faults as they occur and to initiate action that ensures that the minimum of further damage occurs before failure. In its basic form the function of the protective relay is outlined in Figure 1.1.

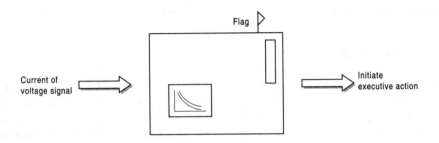

Figure 1.1   The basic function of an electrical protective relay

The signal provided by the transducer will be in the form of a current or voltage and will be interpreted by the relay as an acceptable or unacceptable level, according to a pre-set value determined by the relay designer or the maintenance staff. If the pre-set value is exceeded then the relay will initiate further electromechanical action that will often result in disconnection of the electrical machine, and it will flag the fact that a fault, or even failure, has been identified. This is a simplistic view of the protective relay, which was configured using electromechanical devices such as relays to carry out their function, as the name implies. However, nowadays most protective relays use digital processors to deploy a wide range of functions, and are programmable to allow more sophisticated criteria for initiating interrupt procedures to be applied; for example, to block the restart of a motor until it has cooled to an acceptable degree. Figure 1.2 shows a typical modern programmable relay for fulfilling such a function.

From what has been said earlier it is apparent that protective relaying can be regarded as a form of monitoring, and indeed it is widely used with great success. Modern digital relays have also started to fulfil a monitoring function since they can record the voltages and currents they measure for a period before and after any fault. In fact many failure investigations on electrical machines, involving root cause analysis, start with the download and analysis of the digital protective relay data, which can usually be displayed clearly in an Excel spreadsheet. Virtually all electrical machine protection systems embody some form of protection device, and on typical machines they are used in some or all of the following schemes

- earth fault protection,
- overcurrent protection,
- differential current protection,
- under- and overvoltage protection,

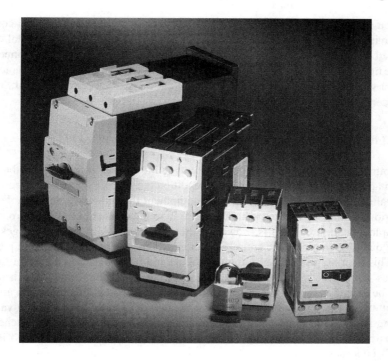

Figure 1.2   A typical modern digital motor relay. [Source: GE Power Systems, USA]

- negative phase sequence protection,
- field failure protection,
- reverse power protection,
- overspeed protection,
- excessive vibration protection,
- thermal overload protection.

This list is representative rather than exhaustive.

It is important to stress the fact that protection is basically designed to act only once a fault has occurred and it will normally initiate some executive action. In the words of Electricity Training Association's *Power Systems Protection* [1], 'the function of protective equipment is not the preventive one its name would imply, in that it takes action only after a fault has occurred; it is the ambulance at the foot of the cliff rather than the fence at the top'. Condition monitoring needs to establish itself as the 'fence at the cliff top'. The executive action may very well be the disconnection of the piece of machinery from the supply. Such action is acceptable if the item of plant is readily dissociated from the process it is involved with, or if it exists substantially in isolation. If, however, the piece of plant is vital to the operation of a process then an unscheduled shutdown of the complete process may occur. The losses involved may then be significantly greater than those resulting simply from the loss of output during a scheduled shutdown. It must also be borne in mind that the capital cost of an individual machine is more often than not small compared with the capital costs involved in a plant shutdown. Maintenance is most effective when it is planned to service many items in the course of a single outage. In summary, condition monitoring of an electrical machine is not necessarily aimed solely at the machine itself, but at the wider health of the process of which it is part.

1.2   The need for monitoring

The notion of the scheduled shutdown or outage introduces us logically to the case to be made on behalf of monitoring. By condition monitoring we mean the continuous evaluation of the health of plant and equipment throughout its serviceable life. Condition monitoring and protection are closely related functions. The approach to the implementation of each is, however, quite different. Also the advantages that accrue due to monitoring are entirely different to those to be expected from protection. This is principally because monitoring should be designed to pre-empt faults, whereas protection is essentially retroactive. Condition monitoring can, in many cases, be extended to provide primary protection, but its real function must always be to attempt to recognise the development of faults at an early stage. Such advanced warning is desirable since it allows maintenance staff greater freedom to schedule outages in the most convenient manner, resulting in lower down time and lower capitalised losses.

We have said that advanced warnings of malfunction, as provided by monitoring, are desirable. Are they? We must justify this because the implementation of a monitoring system can involve the operator in considerable expense. There are other questions to be answered too, for example:

- Once one has chosen to embark upon a programme of monitoring what form should it take?
- Should the monitoring be intermittent, regular at fixed time intervals, or continuous?
- If one employs a fixed time interval maintenance programme then is it necessary to monitor at all?
- Monitoring can generate large quantities of data; how can this information be best used to minimise future expenditure?
- Finally, and perhaps most importantly, how much needs to be spent on monitoring in order to make it truly effective?

These questions do not have simple answers but we can get some indications by considering the magnitude of the maintenance and replacement burden that industry is continually facing, and the implications for the costs of various maintenance strategies. We could consider three different courses of action

- breakdown maintenance,
- fixed-time interval or planned maintenance,
- condition-based maintenance.

Table 1.1   Expenditure on plant per employee of selected
             industries adapted from Neale Report, 1979 [2]

Industry	Annual investment/employee in plant and machinery, £
North Sea oil and gas	160 000
Oil refining	14 000
Electricity supply	8 000
Chemical industry	2 400
Iron and steel	1 800
Water supply	800
Textile manufacture	600
Instrumentation manufacture	400
Electrical engineering manufacture	400

Method (1) demands no more than a 'run it until it breaks then replace it' strategy, while method (2) may or may not include a degree of monitoring to aid in the planning of machinery outages. The final scenario method (3) requires a definite commitment to monitoring.

The scale of investment can be seen from figures provided by the Neale Report [2], published in 1979. This information is 30 years old and comes from a period before a long period of privatisation but is still invaluable. Table 1.1 shows the annual investment per employee in plant and machinery. We have modified these values in order to reflect more realistically today's costs and have selected those industries that would have a high proportion of expenditure in electrical machinery and ancillary plant.

The same report shows that the average annual expenditure on maintenance was 80 per cent of the amount annually invested in plant and machinery. The figures for some selected industries and industrial groupings are shown in Table 1.2, which shows the annual maintenance expenditure as a percentage of the annual plant investment expenditure. This is a high figure in real terms and anything that helps to reduce it must be welcome. The *Hewlett-Packard Journal* has quoted the staggering figure of $200 billion as the annual maintenance bill for US business, and a growth rate of 12 per cent. Now only a fraction of this sum will be spent on maintaining electrical machinery, but even if it amounts to a fraction of one per cent of the total it is still an enormous amount of money.

There are great incentives to maintain plant more efficiently, particularly when it is estimated that approximately 70 per cent of the maintenance work carried out by companies that use no planning at all may be classified as emergency work and must be done at premium costs. It is apparent that careful thought should be given to the most appropriate form of maintenance planning. Breakdown maintenance can only be effective when there is a substantial amount of redundant capacity or spares are available, and a single breakdown does not cause the failure of a complete system.

Table 1.2   *Annual maintenance expenditure as a percentage of annual capital investment in plant, for selected industries. Adapted from the Neale Report, 1979 [2]*

Industry	Maintenance expenditure/plant expenditure, %
Printing	160
Instrumentation manufacture	150
Mechanical engineering	100
Textile manufacture	82
Water supply	80
Gas supply	80
Electricity supply	80
Electrical machinery manufacture	80
Chemical manufacture	78
Marine engineering	50
Iron and steel manufacture	42
Coal production	26

The question to be answered in such circumstances is why is there a significant redundancy? And should it be allowed to continue?

Many sectors of industry, and particularly the electricity, water and gas utilities and the railways, have adopted maintenance planning based on replacement and over-haul at fixed time periods, so that outage work can be scheduled, and diversions and loads can be planned. Such scheduling is usually planned on the basis of plant monitoring, which is typically done on a discontinuous basis. There are many estimates of the savings that accrue by adopting such an approach and an average reduction figure of 60 per cent of the total maintenance burden may be considered reasonable. This is good news, but it must be treated cautiously because such a maintenance policy makes heavy demands upon scarce, skilled manpower. It is also estimated that only 10 per cent of components replaced during fixed-interval maintenance outages actually need to be replaced at that time. The obvious implication is that 90 per cent of what is replaced need not be.

Such considerations, and the realisation that modern electrical machines and the processes they operate in are growing in complexity, leads one to the conclusion that continuous condition monitoring of certain critical items of plant can lead to significant benefits. These benefits accrue in

- greater plant efficiency,
- reduced capitalised losses due to breakdown,
- reduced replacement costs.

The plant operator can also be updated with information on the performance of his machinery. This will help him to improve the day-to-day operational availability and efficiency of the plant. Condition monitoring should give information relevant to both the operational and maintenance functions. There is an added bonus in that better maintenance gives better safety.

In the longer term, condition monitoring also allows the operator to build up a database that can be used for trend analysis, so that further improvements can be made in the scheduling of maintenance. Such information should also be used advantageously by plant manufacturers and designers in order to further improve product reliability. This step effectively closes the loop.

In view of this, how much needs to be spent on monitoring? This depends on the value of the process in which the machine works, and estimates vary, but they are never less than 1 per cent of the capital value of the plant being monitored. A more typical (and probably more realistic) figure would be 5 per cent for the general run of industrial processes, while special requirements for high value processes, such as those found in the offshore oil and gas industry, may push a realistic figure to greater than 10 per cent.

1.3   What and when to monitor

Now that we have examined some of the advantages to be gained from a commitment to condition monitoring we can briefly address the questions, what should we monitor, and when? The question of what to monitor has two implications.

- What machines?
- What parameters?

The first part is more easily answered. In view of the capital costs involved in providing monitoring, whether it takes the form of a permanent installation with its own local intelligence, or a handheld device used periodically by a skilled operator, it is unlikely that electrical machines with ratings of less than 20 kW would benefit. There are, of course, exceptions to this where a smaller machine has a vital function in the performance of a larger system. It will pay dividends to carefully consider the implications of losing the output of an individual piece of machinery in the context of a complete system.

Larger electrical drives, which support generating, process or production plant if a high margin of spare capacity exists, will benefit from monitoring, although perhaps not continuous monitoring. One could include induced and forced-draught boiler fan drives, boiler water feed pump drives, and cooling water pump drives in power stations in this category. It must be borne in mind, however, that successful monitoring can allow a big reduction in the requirement for on-site spare capacity.

Machines that have a high penalty in lost output costs need to be monitored continually. Large generators naturally fall into this category since lost output can exceed £600 000 per day for a large machine in a high-efficiency power station. A similar approach would apply to large propulsion motors and large process drive motors.

The conclusion is that there are machines to which monitoring is readily applicable, but there are other circumstances where careful assessment is needed before deciding. One must always be mindful of the scale of the maintenance burden, however, and not be driven to false economies on the basis that 'nothing has gone wrong so far'. On the other hand one must bear in mind the complexities of the monitoring system itself and its own maintenance burden. Nothing can be worse than investing in complex monitoring equipment, which because of poor design or maintenance gives rise to large numbers of false alarms and leads to the equipment being ignored.

The parameters to be monitored are essentially those that will provide the operator and maintainer with sufficient details to make informed decisions on operation and maintenance scheduling, but which ensure security of plant operation. Automatic on-line monitoring has only recently begun to make an impact in the area of electrical machines. Traditionally quantities, such as line currents and voltages, coolant temperatures, and bearing vibration levels, have been measured and will continue to be used. Other quantities, involving the sensing of pyrolysed products in cooling gases and oils, have recently been introduced, as have techniques for measuring contamination levels in bearing lubricants. Other specialist methods, involving the accurate measurement of rotational speed, or the sensing of leakage fluxes, are being developed in order to monitor a variety of fault conditions.

As the ready availability of sophisticated electronic and microprocessor-based systems is increasingly translated into monitoring hardware, the more variables it is possible to consider, and the more comprehensive the monitoring can be. This trend will be further accelerated as the costs of computing power fall still further, and the complexity of microprocessors increases. Such developments are essential both because of the complexity of the plant being monitored and the complexity of the monitoring signals themselves.

The question of when to monitor is more easily answered. One should monitor when it is cost-effective to do so, or when there are over-riding safety considerations to be observed. The assessment of cost-effectiveness can be a relatively complex matter, but in general terms monitoring is worthwhile when the net annual savings are increased by its use. The net annual saving is the difference between the gross annual saving and the annual costs. The costs of monitoring include the initial investigation, purchase, and installation charges, the staff training costs, and the costs associated with the data acquisition. This expenditure can be written off over the lifetime of the monitoring system and set against the savings accrued. We have already considered these savings in some detail earlier in this chapter, and it is sufficient to say that it is not uncommon for the capital costs of a wisely chosen monitoring system to be retrieved in the first year of its operational life.

Finally it is tempting to think that, with such a degree of monitoring power becoming available, the protective and monitoring functions could be merged. With the development of more powerful digital protection and improved supervisory control and data acquisition (SCADA) systems this is happening but care must be taken and operational experience must be established before these functions merge.

1.4   Scope of the text

Some time ago the authors recognised the need to draw together into a single source an account of the techniques available to anyone wishing to involve themselves in the monitoring of electrical machines.

The list of books on the subject of electrical machine condition monitoring is short, the first historic reference of seminal interest being Walker [3], followed by the present authors' first edition [4] and then by Vas [5]. The most up-to-date book, by Stone *et al.*, is aimed at winding and insulation problems [6].

The journal literature on condition monitoring of electrical machines is growing rapidly. In fact, one author has said that it has picked up at a fervent pace and another has called it an explosion, although the growth is not necessarily in directions most useful to industry. There are a number of general survey papers of condition-monitoring techniques for machines of which the most relevant are Finley and Burke [7], Singh and Al Kazzaz [8], Han and Song [9] and Nandi *et al.* [10].

Rao has given an overview of condition monitoring in his handbook [11], including a chapter on electrical machines, while Barron [12] gives a succinct mechanical engineer's view of condition monitoring which is useful as an overview. It is important in electrical machines monitoring for a bridge to be developed between electrical and mechanical engineers.

Condition monitoring is an area of technology that is extremely wide-ranging, requiring knowledge of

- the construction and performance of the machines to be monitored,
- the way they fail in service,
- the analysis of electrical, magnetic, vibration and chemical signals,
- the design of microprocessor-based instrumentation,
- the processing of these signals and their presentation in a comprehensible way.

In a book of this length it is not possible to enter into a detailed study of each area. The art of condition monitoring is minimalist, to extract the correct information from the machine that enables us, with a minimum of analysis, to give a clear detection of an incipient failure mode.

We have instead set ourselves the objective of covering the complete monitoring field as it relates to electrical machinery, in a manner that will be useful to anyone wishing to become familiar with the subject for the first time, and will assist people actively engaged in condition monitoring to gain a perspective of new developments.

To restrict the field the authors deal with rotating machines only and with techniques that can be applied when those machines are in operation, neglecting the many off-line inspection techniques. We have added two further restrictions. While recognising the enormous growth in recent years of the application of electronic variable speed drives to industry, this edition does not deal with the specific problems of condition-monitoring electrical machines driven at variable speed. In imposing this limit we acknowledge that variable speed drives will be an important area for future growth in condition monitoring, but this technology will be founded firmly upon the behaviour of machines at constant speed, which this book will address. We have also concentrated on conventional machines rather than the emerging brushless, reluctance, permanent magnet and unusual topology machines. This is because the industry is still dominated by these conventional machines and the failure modes for emerging designs are not yet clear, but will be based upon conventional machine experience.

The text is divisible into four sections. The first section, Chapters 2 and 3, is essentially 'a description of the patient, the things that can go wrong with him, and a general guide to the diagnosis'. Chapter 2 gives a broad guide to electrical machine construction with descriptions of the materials and specification limits upon them. It also details examples of faults that can occur and through a number of tables starts to classify the principal failure modes and root causes. Chapter 3 then describes the general principles of reliability theory and its applicability to electrical machines. In particular it highlights the importance of failure modes in predicting failure probability and introduces the idea of condition monitoring addressing the root causes of failure modes that have a slow failure sequence.

The second section, Chapters 4 and 5, gives a detailed account of specific instrumentation and signal processing techniques. We treat instrumentation at the functional level and assume a certain basic knowledge of the techniques of spectral analysis of signals.

The third section, Chapters 6–10, gives a detailed account of specific monitoring techniques, starting with thermal and then chemical degradation methods, progressing onto mechanical and finally electrical methods, considering first terminal conditions and finally discharge monitoring of electrical machine insulation systems. There are areas of overlap between each of the monitoring methods. As far as possible we have subdivided the techniques within each chapter into the types and parts of machines on which they are used. In these sections we describe current practice and discuss some of the new developments now being introduced.

The fourth and final section, Chapters 11 and 12, considers first the application of artificial intelligence to the condition monitoring of machines and then the use of condition monitoring on the maintenance planning and asset management of plant.

We have tried to be mindful of the fact that, when describing developments in a relatively new subject area, a comprehensive bibliography is of the utmost importance. We have in general quoted the major recent journal papers and books that have illuminated the subject. We have only quoted conference papers where they are essential to identify a particularly relevant modern point. We have also included some of the older seminal works, which still deserve scrutiny. It is inevitable that there will be omissions but hopefully it will provide the interested reader with a useful source of additional material.

1.5   References

1.   Electrical Training Association. *Power System Protection, Vol. 1: Principles and Components*. Stevenage: Peter Peregrinus; 1981.

2. Neale N. and Associates. *A Guide to the Condition Monitoring of Machinery*. London: Her Majesty's Stationary Office; 1979.
3. Walker M. *The Diagnosing of Trouble in Electrical Machines*. London: Library Press; 1924.
4. Tavner P.J. and Penman J. *Condition Monitoring of Electrical Machines*. Letchworth: Research Studies Press and John Wiley & Sons; 1987.
5. Vas P. *Parameter Estimation, Condition Monitoring and Diagnosis of Electrical Machines*. Oxford: Clarendon Press; 1996.
6. Stone G.C., Boulter E.A., Culbert I. and Dhirani H. *Electrical Insulation for Rotating Machines, Design, Evaluation, Aging, Testing, and Repair*. New York: Wiley–IEEE Press; 2004.
7. Finley W.R. and Burke R.R. Troubleshooting motor problems. *IEEE Transactions on Industry Applications* 1994, **30**(5): 1383–97.
8. Singh G.K. and Al Kazzaz S.A.S. Induction machine drive condition monitoring and diagnostic research – a survey. *Electric Power Systems Research* 2003, **64**(2): 145–58.
9. Han Y. and Song Y.H. Condition monitoring techniques for electrical equipment – a literature survey. *IEEE Transactions on Power Delivery* 2003 **18**(1): 4–13.
10. Nandi S., Toliyat H.A. and Li X. Condition monitoring and fault diagnosis of electrical motors – a review. *IEEE Transactions on Energy Conversion* 2005, **20**(4): 719–29.
11. Rao B.K.N. *Handbook of Condition Monitoring*. Oxford: Elsevier; 1996.
12. Barron R. *Engineering Condition Monitoring: Practice, Methods and Applications*. Harlow: Longman; 1996.

Chapter 2
Construction, operation and failure modes of electrical machines

2.1 Introduction

This chapter could also be subtitled 'the way rotating electrical machines fail in service'. Rotating electrical machines convert electrical to mechanical energy, or vice versa, and they achieve this by magnetically coupling electrical circuits across an airgap that permits rotational freedom of one of these circuits. Mechanical energy is transmitted into or out of the machine via a drive train that is mechanically connected to one of the electric circuits.

An example of one of the largest electromagnetic energy conversion units in the world, at 1 111 megavolt-amperes (MVA), is shown in Figure 2.1. The construction

Figure 2.1   *View of a 1 111 MVA, 24 kV, 50 Hz steam turbine-driven, hydrogen-cooled, two-pole turbine generator installed in a nuclear power station in the Czech Republic. The generator exciter is on the left, the turbine generator is in the centre of the picture and the low pressure turbine to the right. [Source: Brush Turbogenerators]*

of electrical machines is similar, whether large or small, as shown later in the chapter and their operational weaknesses are dominated by the same principles. The purpose of this chapter is to explain their constructional principles and the main causes of failure. The chapter is illustrated with a large number of photographs to demonstrate to the reader the salient features of electrical machines.

2.2   Materials and temperature

The magnetic and electric circuits essential to machines require materials of high permeability and low resistivity, respectively, and these are generally metals. Metals with good magnetic and electrical properties do not necessarily have high mechanical strength. Indeed the atomic structure of a good conductor is such that it will naturally have a low yield strength and high ductility. Yet the magnetic and electric circuits of the machine must bear the mechanical loads imposed upon them by the transfer of energy across the airgap. Furthermore, the magnetic and electrical circuits must be separated by insulating materials, such as films, fibres and resins, which have even weaker mechanical properties. Table 2.1 sets out the elastic moduli and tensile strength of materials used in electrical machines and highlights the relative weakness of electrical steel, conductor and insulating materials. Right from the outset then, there is a conflict between the electrical and mechanical requirements of the various parts of an electrical machine, which the designer must attempt to resolve.

Table 2.1   Mechanical properties of materials used in electrical machines

Material	Elastic modulus, GPa	Tensile strength, MPa
High tensile steel	210	1 800
Structural steel	210	290–830
Electrical steel	220	450
Copper	120	210
Aluminium	70	310
Epoxy-mica-glass composite	60	275
Moulded organic/inorganic resin	5	48
Phenol-formaldehyde resins	3	35

However, there is a further complication. The transfer of energy inevitably involves the dissipation of heat, by ohmic losses in the electric circuit and by eddy current and hysteresis losses in the magnetic circuit. The performance of the insulating materials that keep these circuits apart is highly dependent upon temperature, and deteriorates rapidly at higher temperatures. Materials that can sustain these higher temperatures become progressively more expensive and their mechanical and dielectric properties are often worse than lower temperature materials. Table 2.2 classifies

Table 2.2   Temperature capabilities of insulating materials

Class	Material	Temperature rating to give an acceptable life under prescribed industrial conditions, °C
O or Y obsolete	Oleo-resinous natural fibre materials, cotton, silk, paper, wood without impregnation.	90
A	Natural fibre materials, cotton, silk, paper and wood impregnated, coated or immersed in dielectric liquid, such as oil.	105
E	Synthetic-resin impregnated or enamelled wire not containing fibrous materials such as cotton, silk or paper but including phenolics, alkyds and leatheroid.	120
B	Combinations of mica, glass and paper with natural organic bonding, impregnating or coating substances including shellac, bitumen and polyester resins.	130
F	Combinations of mica, glass, film and paper with synthetic inorganic bonding, impregnating or coating substances including epoxy and polyester resins.	155
H	Combinations mica, paper, glass or asbestos with synthetic bonding, impregnating or coating substances including epoxy, polymide and silicone resins.	180
C	Combinations of asbestos, mica, glass, porcelain, quartz or other silicates with or without a high temperature synthetic bonding, impregnating or coating substance including silicone. These can include high-temperature aramid calendared papers like Nomex.	220

the common insulating materials used in electrical machines and shows the relatively low temperatures at which they are permitted to operate.

Uncertainties about the temperatures within a machine mean that the designer is forced to restrict the maximum measurable operating temperature to an even lower value than that given in Table 2.2, taken from the IEC standard [1], for the appropriate insulation, in order to provide a safety factor during operation. It is clear that the heat dissipated within a machine must be removed effectively if design limits are to be met. For example, in the 1 111 MVA turbine generator shown in Figure 2.1 with losses of the order of 12 MW, if cooling stopped the average temperature of the generator body would exceed any of the maximum permitted insulation temperatures within 12 seconds. The problem is exacerbated because the losses are not evenly distributed and in practice at some locations the rise in temperature will be even faster than this. So cooling and its distribution become a vital part of machine design.

The health of an electrical machine, its failure modes and root causes, are ultimately related to the materials of which it is made, the mechanical and electrical stresses those materials are subjected to and the temperatures they attain in service.

In Chapter 1 we explained how electrical machines are protected by relays, which sense serious disruptions of the current flowing in the windings and operate to trip or disconnect the machine. However, when fault currents are flowing the machine has already failed as an electrical device. Electrical or mechanical failure modes are always preceded by deterioration of one of the mechanical, electrical, magnetic, insulation or cooling components of the machine. This is the case regardless of the type of electrical machine. If this deterioration takes a significant period of time and can be detected by measurement, then that root cause detection will be a means of monitoring the machine before a failure mode develops. The heart of condition monitoring is to derive methods to measure, as directly as possible, parameters that indicate root cause deterioration activity and provide sufficient warning of impending failure in order that the machine may be taken off for repair or may be tripped before serious damage occurs.

A degree of protection could be achieved by making the protective relays especially sensitive and providing an alarm indication before tripping occurs. Experience has shown that this is a precarious mode of condition monitoring leading to false alarms and a lack of confidence in the monitoring process. The following sections show how the construction, specification, operation and types of fault can lead to the identification of generic failure mode root causes in the machine.

2.3   Construction of electrical machines

2.3.1   General

The basic constructional features of the electrical machine are shown in Figure 2.2(a). The rotor, which usually has a relatively high inertia, is normally supported on two bearings, which may be mounted on separate pedestals or incorporated into the enclosure of the machine, as shown in Figure 2.2(a). Some larger, slower-speed machines incorporate a single non-drive end bearing and rely on the prime mover or driven plant and its bearings for the remaining support. Rolling element bearings are used

(a)

(b)

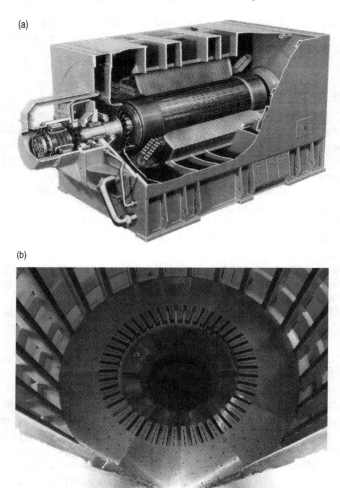

Figure 2.2   *(a) Medium-sized synchronous generator. Section through a 125 MVA, 15 kV, 60 Hz, two-pole, air-cooled, brushless excitation turbine generator showing the fabricated main frame of the machine, stator core, winding, rotor and on the right the main exciter of the machine. (b) Large synchronous generator. Construction of the stator core of a 500 MW, two-pole, hydrogen-cooled turbine generator showing the fabricated inner frame of the machine and the segmented laminations being inserted into that frame. [Source: Brush Turbogenerators]*

on smaller-size machines where shaft peripheral velocities are low, and sleeve bearings with hydrodynamic oil films are used for larger machines. Vertically mounted machines will incorporate a thrust bearing usually at the low end of the enclosure. This may be a relatively modest angular contact ball bearing for a small, vertically mounted pump motor but could be a large hydrodynamic oil film thrust pad Michell-type bearing for a hydro-type generator where the rotor may weigh 100 tonnes or more (see Figure 2.6(b)).

2.3.2   Stator core and frame

The stators of all AC machines are constructed from lightly insulated laminations of electrical steel. As Table 2.1 shows, electrical steels are strong but the silicon, incorporated into the alloy, to raise the resistance and impart magnetic properties, weakens the material compared to structural steel, making it brittle. Furthermore, if the laminated structure is to have the cohesion necessary to transmit the load torque, and have low levels of vibration when carrying the magnetic flux, it must be firmly clamped between cast or fabricated end-plates that are secured to a cylindrical frame into which the core is keyed. The core is constructed within the frame and compressed before the clamping plates are applied. The frame structure and its clamping are clearly visible in Figures 2.2(b), 2.3(a) and 2.5(b). On larger machines the clamping plates are tightened by large bolts (see Figure 2.3(a)), but on smaller machines interlocking keys or even welds are used to secure the clamping plates, or the core itself may be welded or cleated.

In a DC machine the laminated stator field poles are bolted to a rolled-steel yoke that has much greater inherent strength than a laminated core (Figure 2.7(a)).

2.3.3   Rotors

The design of the rotor will depend on the particular type of machine. AC induction and DC motors have laminated rotors where the laminations are clamped together and shrunk onto the steel shaft (Figure 2.7(a)). Turbine-type generators have large, solid, forged-steel rotors that are long and thin (Figure 2.3(b)), while hydro-type generators have large, short, fat rotors with laminated pole shoes bolted onto a fabricated spider (Figure 2.6(b)). Where air or gas cooling is necessary an axial or radial fan may be fitted at either or both ends of the rotor shaft. However, smaller machines rely solely on air circulation as a result of the windage of the rotor itself, which is usually slotted to accept the rotor windings (Figure 2.9).

The rotor windings of generators are constructed of hard-drawn copper and are insulated with rigid epoxy or formaldehyde resin and impregnated into a woven material. On squirrel cage induction motors the winding may consist of lightly insulated copper bars driven into the slots in the laminated rotor or of aluminium bars cast directly into the rotor. The rotor windings of a DC machine or wound rotor induction motor are rather similar to a conventional AC stator winding that is described later. Typical induction motor and generator rotors are shown in Figures 2.3 and 2.10 .

2.3.4   Windings

The stator windings of all high-voltage AC machines comprise conductor bars made up of hard-drawn, higher strength copper subconductors that may be connected in

(a)

(b)

Figure 2.3   *(a) Large synchronous generator. Stators for 2 500 MW, two-pole, hydrogen-cooled turbine generators. The stator nearest the camera is wound. The stator furthest from the camera is awaiting winding. (b) Large synchronous generator. Rotor for a 500 MW, two-pole, hydrogen-cooled turbine generator showing rotor forging and rotor winding before the fitting of end bells. [Source: Brush Turbogenerators]*

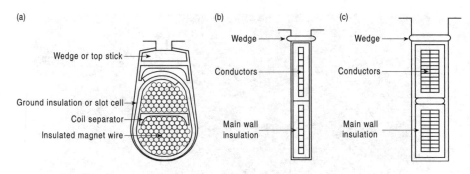

Figure 2.4   Sectional view of the slot section three stator windings, not to the same scale. (a) A 400 V round wire winding for a 1 kW motor. (b) An 11 kV, 1.5 MW motor winding. (c) A 23 kV conductor bar for the stator winding of a large turbine generator

series or parallel. Individual subconductors are covered with a paper or glass-based tape and the assembled bar is overtaped with a similar material impregnated on older designs with bitumen but nowadays with epoxy resins (see Figure 2.4). In the portion of the conductor bar embedded in the stator slot the insulation system is compacted by being heated and pressed or it may be impregnated under vacuum and pressure. In the end-winding portion, where one coil is connected to another, the insulation system is not compacted and may be slightly altered, containing less impregnant, so that it is more flexible and therefore better able to withstand the large electromagnetic forces that part of the winding experiences. An important part of the construction is the manner of the bracing of these end windings. They are usually pulled back onto rigid insulated brackets made of impregnated laminate or steel using nylon or Terylene lacing cord. On the largest machines (Figure 2.5(a)) bracing rings of glass-reinforced plastic are used with insulating bolts. The exact nature of the bracing depends upon the machine rating and the relative length of the end winding, as determined by the number of pole pairs. The yoke (or stator core) is fitted into a frame and enclosure. On smaller machines and those of standard design, the stator core is secured directly into a simplified design of a machine main frame (Figure 2.10), but on larger machines the core has its own inner frame, which is separate from the outer frame so that the clamped core can be removed from the enclosure for repair (Figure 2.5(b)).

2.3.5   Enclosures

The machine enclosure can take a wide variety of forms, depending on the manner in which the machine is cooled, and the protection it needs from the environment in which it will work. Where a pressurised gas system of cooling is used the enclosure will be a thick-walled pressure vessel but for simple air-cooling with an open-air circuit the enclosure will consist of thin-walled ducting. Typical enclosures are shown

(a)

(b)

Figure 2.5   *(a) Large synchronous generator. End region of a 600 MW, two-pole,*
hydrogen-cooled turbine generator with water-cooled stator windings
showing the end winding bracing structure and the hoses carrying water
to the winding. (b) Large synchronous generator. Stator of a 600 MW,
two-pole, hydrogen-cooled turbine generator showing stator core, frame
and windings being inserted into its stator pressure housing preparatory
to factory testing. [Source: Dongfang Electrical Machinery, China]

(a)

(b)

Figure 2.6   *(a) Large synchronous generators. Generator Hall in the Grand Coulee Hydroelectric Dam showing a number of hydro generator units, 120 MVA, 88-pole, 60 Hz, 81.8 rev/min. [Source: Grand Coulee Dam, USA] (b) Large synchronous generator. Installation of the stator of a 75 MVA, 44-pole, 50 Hz, 136.4 rev/min hydro generator in Iceland showing stator core, frame and windings being lowered over the 44-pole rotor supported on its thrust bearing. [Source: Brush Turbogenerators]*

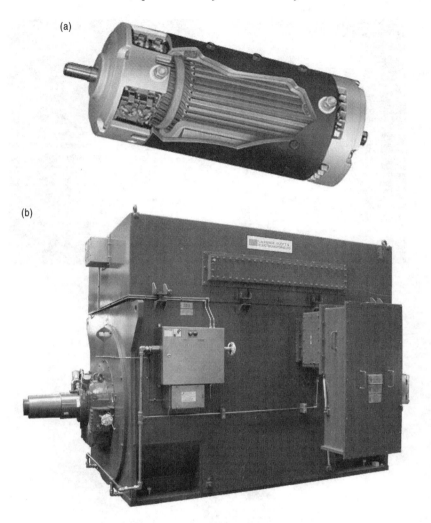

Figure 2.7   *(a) Small DC motor. Section through a 500 W, wound-field DC motor showing stator frame and bearing housings, armature and commutator on the left. [Source: SEM Motors, UK] (b) Large AC induction motor. 7 MW, 11 kV, 50 Hz, four-pole, 1486 rev/min designed to drive a high-speed flash gas compressor for offshore oil and gas. Note the drive shaft on left and the water-cooled heat exchanger on top of machine. [Source: Laurence, Scott & Electromotors Ltd.]*

in Figures 2.1, 2.2(a), 2.5(b), 2.7, 2.8 and 2.9. There is an increasing demand nowadays to reduce the noise level from electrical machines and apart from affecting the basic design of the stator and rotor cores, this will require specially designed noise-proof enclosures.

(a)

(b)

Figure 2.8   *(a) Large AC induction motor. View of a 20 MW advanced induction motor for ship propulsion showing low speed drive shaft and heat exchangers on the machine flanks. This is a large multi-phase, multi-pole, variable speed motor fed by a current-fed inverter. Source: Alstom, France. (b) Small AC motor. View of a combined 4 kW motor and inverter used in a small Nissan full-electric-vehicle, the Hyper-Mini. [Source: Hitachi, Japan]*

(a)

(b)

Figure 2.9   *(a) Medium synchronous generator. 400 V line, 40 kVA, four-pole generator for diesel genset, pilot exciter on the right. (b) Small AC motor. Induction motor, 400 V line, 40 kW. [Source: Marelli Motori S.p.A.]*

2.3.6   Connections

Electrical connections are made to the windings via copper busbars or cables that leave the machine enclosure through bushings into a terminal box. The main three phase busbars of the 1 111 MVA generator are visible rising from the stator frame in the centre of Figure 2.1. The busbars may be lightly insulated to protect them against the environment. The bushings usually consist of the busbar embedded into an epoxy resin casting, although wound paper bushings may be used on older machines. The electrical connections are well braced to withstand the large electromagnetic forces that are developed when fault currents flow. The terminal enclosure allows the proper termination of the supply cables or busbars, and must be specially designed to suit the environment in which the machine works. For example, special enclosures are required for motors that operate in inflammable areas and these incorporate baffles and seals to ensure that any flashover in the enclosure does not ignite gas or vapour outside the terminal box.

Many machines incorporate brushgear for connection to the rotor windings either through steel or copper sliprings or through a copper commutator (Figure 2.7(a). The commutator is a very carefully designed component in which copper segments interlock with the rotor so that they can withstand the bursting forces acting upon them. Also, each segment must be well insulated from its neighbours, and mica is normally used for this purpose. Sliprings are usually shrunk onto an insulating sleeve mounted on a boss on the rotor shaft, and electrical connections to the sliprings are insulated and carefully braced to withstand the centrifugal forces upon them. Brushes are springloaded and mounted in brass brush boxes around the periphery of the rings or commutator.

Heat exchangers for the cooling system of the machine are mounted on the enclosure or may be a part of it, as shown in Figure 2.7(b). They may be as simple as a finned casing to the machine to promote convective heat transfer to the surrounding air or they may be a more complex water-cooled system through which the cooling gas or air is ducted.

2.3.7   Summary

These descriptions show the very wide range of materials that are used in an electrical machine and Table 2.3 gives a summary of these based on the structure of the machine. In particular it should be noted how important insulating materials are, both in terms of volume and cost in the overall structure of an electrical machine. In the following section this structure of the machine will be discussed in more detail.

2.4   Structure of electrical machines and their types

The previous sections provided a brief description of the major constructional components of an electrical machine and the materials of which they are made. The difference between the structure for assembly and for reliability will be described in Chapter 3.

Table 2.3   Materials used in the construction of a typical electrical machine

Subassembly	Component	Materials
Enclosure	Enclosure Heat exchanger electrical connections Bushings bearings	Fabricated structural steel Steel, copper or brass tube Copper or aluminium busbar or cable Cast epoxy resin Steel babbitt, high tensile steel rolling elements or soft bearing alloy on bearing shells
Stator body	Frame Core Core clamp	Structural steel Electrical steel laminations or rolled steel yoke Structural steel or non-magnetic, low-conductivity alloy
Stator winding	Conductors insulation End winding support	Hard drawn copper or copper wire Mica-paper or glass or film impregnated with resin Glass fibre structural materials and impregnated insulation felts, ropes and boards
Rotor winding	Conductors Insulation End winding support	Hard drawn copper or copper wire Mica-paper or glass or film impregnated with resin Impregnated glass fibre rope
Rotor body	Shaft Core Core clamp slip rings Brushgear	Structural steel or forging Electrical steel laminations or steel forging integral with shaft Structural steel or non-magnetic, low conductivity alloy steel, brass or copper Carbon or copper brushes in brass brushholders

In this section the detailed structure for assembly of the electrical machine is discussed and the effect of different types of machine upon it. This structure is exemplified by Figure 2.10. Note the similarity between this 4 kW induction motor and the 125 MVA synchronous generator shown in Figure 2.2(a).

This structure is also presented in tabular form in Table 2.3 and this will form the basis for considering the faults in machines later in the chapter. Note that generators require an exciter to provide the field current for their rotor. They generally have their exciters mounted on the shaft of the main machine and a large generator can have a pilot exciter and main exciter. The main exciter is clearly visible on the left in

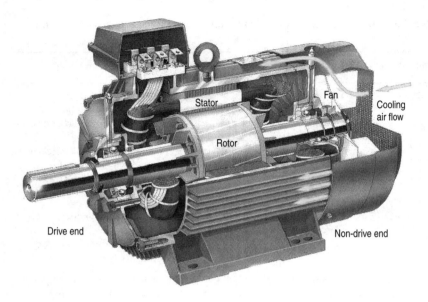

Figure 2.10   Structure of an electrical machine showing stator core and windings, stator frame rotor and bearings of a 4 kW induction motor. [Source: GE Power Systems, USA]

Figures 2.1 and 2.2(a), and the pilot exciter is also visible in Figure 2.2(a) between the main exciter and the main generator. Therefore a generator may consist of two or even three electrical machines on the same shaft. This is important when carrying out a failure modes and effects analysis, in considering the number of subassemblies in the machine (see Chapter 3), when the number of components affects the predicted failure rate of the machine.

Deterioration of performance or failure in service can occur due to damage of any of these components, and the descriptions of failures at the end of this chapter show how wide-ranging these root causes can be. However, experience shows that particular components are under specific electrical, mechanical or thermal stresses and to a large extent these components depend on the size and type of the machine.

Table 2.4 gives a range of typical sizes for some larger electrical machines and shows some important mechanical, electrical and thermal parameters in the design of electrical machines, which determine their ability to meet the conditions described in the previous sections, namely

- magnetic loading, T,
- electric loading, A/mm,
- current density, A/mm^2,
- mechanical loading, peripheral velocity, mm/s.

The values of these parameters in Table 2.4 give an idea of the limits to these parameters in all electrical machines. Table 2.5 sets out the major types of electrical machine,

Table 2.4   Sizes and properties of some typical large electrical machines. Taken from Tavner and Anderson [2]

Type of machine	Air-cooled induction motor	Air-cooled turbine generator	H_2-cooled turbine generator	Air-cooled hydro generator	H_2O and H_2-cooled turbine generator
Stator winding cooling	Air indirect	Air indirect	H_2 indirect	Air indirect	H_2O direct
Stator core cooling	Air direct	Air direct	H_2 direct	Air direct	H_2 direct
Rotor cooling	Air direct	Air direct	H_2 direct	Air direct	H_2 direct
S (MVA)	7.11	125	108.6	100	1 111
P (MW)	6.14	100	92.3	90	1 000
Poles	4	2	2	88	2
Rev/min	1 790	3 000	3 000	68.2	3 000
Core length, mm	628	4 450	2 800	1 280	7 200
Rotor diameter, mm	742	890	924	12 968	1 200
Typical weight, tonnes	18	90	120	400	350
Mechanical loading, rotor peripheral velocity (m/s)	69.5	139.8	145.1	46.3	188.5
U (kV)	6.8	15	13.8	15.75	24
Magnetic loading, air gap flux density (Trms)	0.9	0.94	0.83	1.05	1
I (A)	604	6 560	4 543	3 666	26 729
Electric loading (A/mm)	93.0	116.3	146.8	62.5	255.3
Current density, (A/mm^2)	3.9	2.8	3.8	3.7	11.1

Table 2.5 Types of rotating electrical machine

Main type of machine	Subtypes	Size range	Main constructional features	Root causes
Synchronous machines	Steam and gas-turbine driven generator	c. 25–1 500 MVA	Length ≫ diameter Horizontally mounted High rotational speed, 1 500–3 600 rev/min Close circuit air or gas-cooling Water cooled stator windings in larger sizes Fossil fuel and nuclear generation	Electrical faults incur consequential damage High-speed balance problems Excitation faults exacerbate balance problems Stator end winding bracing problems Rotor end bell integrity problems Gas-cooling circuit sealing problems Water-cooling circuit sealing problems High performance oil film bearings Enclosed casing means difficult to inspect frequently
	Water turbine-driven generator	c. 10–800 MVA	Diameter ≫ length Vertically or horizontally mounted Low rotational speed, 80–300 rev/min Air-cooled Water-cooled stator windings in larger sizes Hydroelectric generation	Electrical faults incur consequential damage Air-cooling and high stresses lead to discharge erosion on high-voltage windings Rotor pole integrity problems High-performance oil film thrust bearing Enclosed casing means difficult to inspect frequently
	Engine-driven generator	c. 10 kVA–60 MVA	Length ≈ diameter Horizontally mounted Medium rotational speed 500–1800 rev/min Standby, CHP, island and marine generation	Electrical faults incur consequential damage Reciprocating engines give torsional and seismic vibration problems Arduous environment can cause problems

	Synchronous motor	1–30 MW	Length ≈ diameter Horizontally mounted Wide range of possible rotational speeds 500–6 000 rev/min North American technology Reciprocating and turbo-compressor drives	High-speed compressors give high speed balance problems Reciprocating compressors give torsional and seismic vibration problems Stator end winding bracing problems
Permanent magnet machines	Synchronous motor	0.5–500 kW	Length ≫ diameter Horizontally mounted High rotational speed, 500–5 000 rev/min Variable speed drive applications	Electrical faults incur consequential damage High speed balance problems
	Synchronous generator	5–5 MW	Diameter ≫ length Vertically or horizontally mounted Low rotational speed, 80–300 rev/min Renewable applications	Electrical faults incur consequential damage
Induction machines	Generators	100 kVA–10 MVA	Horizontally mounted Medium rotational speeds 500–3 600 rev/min Becoming common in wind generation	Short airgap dives airgap stability problems limiting permissible bearing wear particular in high poleage low speed machines Slip ring and brush gear problems on doubly fed machines
	Motors	0.1 kW–20 MW	Length ≈ diameter Horizontally mounted With variable speed drives wide range of possible rotational speeds 500–6 000 rev/min Ubiquitous industrial workhorse	High starting current limits thermal and fatigue life Short air-gap dives air-gap stability problems limiting permissible bearing wear particular in high poleage low speed machines Rotor cage integrity after many restarts Slip ring and brush gear problems on doubly fed machines

Continued

Table 2.5 Continued

Main type of machine	Subtypes	Size range	Main constructional features	Root causes
DC machines	Generators	20 kW–5 MW	Length ≈ diameter Horizontally mounted Medium rotational speed 500–1 800 rev/min Declining application generally for traction and special applications	Commutator faults Brushgear faults Carbon dust contamination problems Associated control gear problems
	Motors	1 kW–2 MW	Length ≈ diameter Horizontally mounted Medium rotational speed 500–1 800 rev/min Declining application to specialised variable speed applications particularly traction	Intermittent duty thermal problems High shock loading problems Commutator faults Brushgear faults Carbon dust contamination problems Associated control gear problems
AC reluctance machines	Motors	1–50 kW	Length ≈ diameter Horizontally mounted Medium rotational speed 500–1 800 rev/min Increasing application to specialised variable speed high torque applications	Vibration Stator winding damage Variable speed drive problems
AC commutator machines	Motors	20 kW–5 MW	Now almost obsolete in the face of power electronic variable speed drives applied to induction and DC motors	Commutator faults Brushgear faults Carbon dust contamination problems Associated control gear problems

which shows the main constructional features of each type and the factors most likely to lead to faults.

2.5 Machine specification and failure modes

Many faults occur because machines are incorrectly specified for the application to which they are being applied. For example, a machine may be underpowered or have an inadequate enclosure. The specification of a machine must ensure that it is of an appropriate design for the use to which it is being put. It is a waste of time applying sophisticated monitoring techniques to a machine that is unfit for its purpose. It is far better to remove the monitoring and change the machine for one that is more suited to the application. By the same token many operational problems could be avoided by using an over-designed machine. For example, in a hot environment it may be better to use an over-rated machine, which has a substantial design margin, than push an adequately designed machine to its limit. On the other hand, it is sometimes an operational fact of life, especially with an expensive machine, that it must continue to operate even though it suffers from shortcomings in the original specification. In such cases effective monitoring can help to ease the burden placed upon the maintenance engineer.

The specification of a machine must reflect the mechanical, electrical and environmental conditions in which the machine will work. These matters will have a bearing on the mechanisms by which the machine may fail in service. The need for monitoring and the selection of the parameters to be monitored must be affected by these operational conditions. Table 2.6 sets out the operational conditions that are covered by a specification and that are relevant to monitoring. These operational conditions are described by Bone and Schwarz [3].

Mechanically, machines can be exposed to periods of intermittent running, frequent starting and to arduous duty cycles, where the load varies frequently between no-load and full-load with occasional overloads. These can lead to insulation degradation, bearing wear, vibration and slackening of windings, commutator or brushgear. Similarly a machine driving a pulsating load, such as a compressor, is going to experience heavy bearing wear, vibration and slackening of windings, commutator or brushgear.

From an electrical supply point of view a machine, by virtue of its location in a supply system or its task in a manufacturing process, may be subjected to a variety of transients at its supply terminals. These may be slow fluctuations in the supply voltage or even unbalance between the three phases that can cause operational problems, for example, if the machine does not have the thermal capacity to deal with the overheating that unbalance can lead to. More rapid transients in the supply voltage, however, can overstress the winding insulation because the electric stress is not uniformly distributed throughout the winding length. Modern interrupters produce very rapid voltage surges that were known to break down class B (see Table 2.2) inter-turn insulation on the line end coils of motors but with modern class F systems this is now rare. The most severe electrical transients a machine can receive, however, are

Table 2.6   Operational conditions, defined by specification, which affect failure mechanisms

Operational condition	Nature of condition	Detailed condition	Root causes and failure modes
Mechanical	Characteristics of the load or prime mover	Duty cycle Pulsating load repeated: starting load or drive vibration	Successive overloads may cause overheating or bearing damage. May cause bearing and low cycle fatigue damage Repeated application of high starting torques may cause excessive overheating and damage to rotor and stator end windings May cause high cycle fatigue damage
Electrical	Characteristics of the electrical system and the machine connected to it	Slow voltage fluctuation Fast voltage fluctuation System unbalance	May cause loss of power and motor stall or generator pole-slip May disrupt generator excitation or cause insulation failure in winding May cause stator winding heating
Environmental	Characteristics of the environment in which the machine is being used	Temperature Humidity Cleanliness	High ambient temperature will cause faster insulation degradation and bearing deterioration. Low ambient temperature may cause frosting and icing High humidity may lead to condensation and tracking damage to insulator surfaces and corrosion of metal parts leading to susceptibility to corrosion fatigue failure Dirt from the environment may enter the machine, contaminating insulation surfaces, enhancing the propensity to tracking, fouling heat exchangers and contaminating brushgear

during starting or reswitching of the supply, and part of the duty of many machines in industrial processes is to be repeatedly started and run for short periods. This will cause overheating, slackening of winding systems, movement of electrical connections and overstressing of terminal boxes.

Environmentally, there are thermal and contamination problems. The machine may run exceptionally hot, because of cooling problems, ambient conditions or simply that the machine is being operated to its rating limit. These can deteriorate its insulating materials. The machine may be operating in a dirty environment either because of the industrial process in which it is working, such as a textile or paper works, or because it has brushgear that produces carbon dust. If dirt can enter the main coolant circuit it may contaminate windings, bushings and electrical connections causing a deterioration in insulation integrity. Alternatively, it may foul coolers, seals, or bearings causing overheating and mechanical damage. The cooling gas may also become damp because of ambient conditions, for instance in a tropical country, or due to cooler leakage. Either of these can lead to the condensation of moisture on the electrical insulation and connections giving a reduced insulation resistance. A machine needs to be designed to meet the environmental, mechanical and electrical disturbances it is likely to encounter during its life but any monitoring scheme that is installed should be directed towards detecting the untoward effects of these disturbances.

2.6 Insulation ageing mechanisms

2.6.1 General

Electrical insulation faults are a significant contributor to the failure of rotating machines. Industry studies described in the next chapter indicate that up to one-third of rotating machine failures can be attributed to loss of function of the stator winding electrical insulation. Therefore they deserve special consideration before we consider failure modes in general. However, it should be borne in mind that although the final failure mode may be electrical breakdown of a dielectric component, the underlying mechanism driving the breakdown may be thermal, mechanical or environmental stress as well as electrical factors. This section will cover the basic stresses that affect the performance of stator winding, stator core and rotor winding insulation systems on operating machines as well as discussing the roles that design, operation and maintenance have on the life of the equipment.

Insulation in service is exposed to high temperature, high voltage, vibration and other mechanical forces, as well as some adverse environmental conditions. These stresses can act together or individually to degrade insulation materials or systems. Thermal ageing of insulating material due to high temperatures has been studied the most and is perhaps best understood. The mechanism may be treated as a chemical rate phenomenon, described by the Arrhenius relationship, and includes loss of volatiles, oxidation, depolymerisation, shrinkage and embrittlement. In actual service, loss of insulation system integrity is aggravated by cyclic and transient mechanical forces, which cause relative movement and abrasion of insulation. Furthermore, insulation subjected to high voltage can degrade due to partial discharge activity. Eventually, the stresses will so weaken the insulation that puncture results and the conductor is connected to earth. Thus, although the final result is electrical failure, the root cause may be the result of non-electrical stress. In general, the higher the electrical stress, the more rapidly the insulation will age. In the discussion to follow, each of these stresses, thermal, electrical, mechanical and ambient conditions will be described in generic terms. The exact details of which mechanisms are the most critical are very much dependent on the type of equipment and service conditions.

2.6.2　Thermal ageing

Thermal ageing occurs when the temperature of the insulation is high enough to cause the electrical and mechanical properties of the insulation to degrade. Cycling of the temperature can also induce mechanical stresses causing deterioration, even if the temperature alone is insufficient to cause damage; for example, the loss of the copper/insulation bond in the stator windings of rotating machines as a result of successive heating and cooling of a conductor.

The operating temperatures of inorganic insulating materials (see Table 2.2) – for example, porcelains and glasses – are limited by softening, by reversible changes of conduction, dielectric loss or strength, or by the danger of fracture due to differential thermal stresses. Organic materials suffer irreversible changes at high temperatures. Generally, the temperature limit that will restrict deterioration to what is acceptable over the design life of the equipment is lower than that imposed by immediate changes such as softening, except when the temperature rise in service is of short duration. Typical obvious symptoms are shrinkage, hardening, spontaneous cracking or crazing, loss of strength, embrittlement, discolouration, distortion and, in extreme cases, charring. These effects are generally due to, or accompanied by

- loss of weight resulting from evaporation of volatile components,
- oxidation or pyrolysis to form volatile substances or gases such as CO, CO_2, water and low-molecular-weight hydrocarbons,
- excessive cross-linking.

2.6.3　Electrical ageing

2.6.3.1　General

Electrical ageing occurs when the electric stress applied to insulation causes deterioration. Although electric stress due to DC and transient voltages can cause ageing, AC voltage is normally the most severe. It should be noted that the insulating materials used in practical equipment operate well below their inherent breakdown strength. Consequently, electrical ageing of insulation usually occurs as the result of the presence of faults in the material; for example, gas voids arising from imperfect impregnation, resulting in partial discharge. The following are electrical ageing mechanisms that can be induced by the principal power frequency voltage or by the transient surge voltage from power system disturbances.

2.6.3.2 Partial discharges

In general, deterioration from partial discharges will occur in insulation that has voids created during manufacture or by thermal or mechanical ageing in service. The direct impingement of discharges on insulation surfaces will cause decomposition of the solid insulation.

2.6.3.3 Surface tracking and moisture absorption

Electrical tracking; that is, the formation of conductive or carbonised paths over an insulation surface, is caused by AC electrical stress. When an insulating surface collects dust and moisture it can become conducting. The conduction current can become large enough to swamp the effects of capacitance currents so that the voltage distribution is no longer determined by capacitances, as for a clean surface, but depends on how the conductance varies over the surface. The pollution film is never uniform, and conduction through the film causes drying out that is most rapid in the regions of highest resistivity. The drying out causes a further increase in resistivity in these regions, so the effect is cumulative and eventually a dry band is formed across the insulation. Most of the voltage across the surface appears across this dry band and flashover can occur over the dry band, constituting a partial breakdown. An arc is formed in the air adjacent to the dry band, which can do one of three things: extinguish rapidly, stabilise and continue to burn, or extend in length until it bridges the insulation causing complete breakdown. In the final stages the arc may grow slowly by thermal drying, which is a reversible process, or may propagate rapidly and irreversibly due to the stress concentration at the arc root. Arcs on polluted insulation can propagate and cause failure at working stresses of 0.02–0.04 kV/mm. These stresses should be compared with average breakdown stress of 3 kV/mm for uniform fields and 5 kV/mm for point-plane gaps in air.

2.6.3.4 Transient voltages

Transient or surge voltages can result from a number of causes such as lightning strikes, switching operations, faults or from the power electronic devices feeding the machine in the case of adjustable speed drives. Normally, insulation systems are designed and qualified to withstand lightning and switching surges; however, faster transients, such as those associated with drives or gas insulated switchgear operations, have been known to cause failures due to the non-uniform voltage distribution resulting from steep-fronted surges. Consequently, part of the design process requires knowledge of the electrical environment and the surge environment in which the equipment will be operating. Transient events can also cause failure in older equipment that has been subject to several years of thermal, electrical, mechanical or ambient ageing.

2.6.4 Mechanical ageing

Mechanical stress is a major direct or indirect contributor to many insulation system ageing mechanisms. Mechanical ageing is principally produced by relative motion of insulating components and results from mechanical and electromagnetic forces, resonances, inadequate wedging and bracing, flexing, abrasion and environmental factors that affect the mechanical properties of the insulation.

2.6.5 Environmental ageing

The contamination of insulation systems by water, oils or chemicals can cause insulating materials to degrade mechanically and electrically. Thermal ageing will also proceed at a faster rate in air than in relatively inert atmospheres such as hydrogen. The effects of moisture or water absorption are well known. The most common ageing effect is the worsening of the electrical properties of the insulation material and a general tendency to poorer mechanical performance. In practice, the severity of problems, caused by the presence of moisture in the insulation, is dependent on the type of insulation system. Older insulation systems containing organic binders and bonding varnishes are more susceptible to mechanical degradation from water absorption than modern systems containing inorganic binders and synthetic resins such as epoxy. However, polyester-based materials can lose their electrical and mechanical properties in wet environments. Hydrolysis is a mechanism by which moisture causes rupture of the chemical bonds of the insulation. This tends to cause delamination and swelling of the insulation exposing it to the risk of failure due to thermal, electrical or mechanical factors. Oils, acids, alkalis and solvents can also attack certain insulating materials and bonding agents. Again, organic materials are more likely to suffer significant deterioration. Contamination of insulating materials with dust can result in electrical failure due to surface tracking.

There are a number of high-voltage applications, such as primary heat transport pump motors, within the reactor containment area of nuclear generating stations, where radiation levels can be high. If acceptable insulation life is to be achieved, the insulation systems used must contain materials with a high radiation resistance to prevent rapid mechanical deterioration of mechanical properties. Materials such as ceramics, mica and glass are known to be only slightly affected by the radiation levels encountered in these areas. Organic materials, on the other hand, are strongly affected by ionising radiation while polymers with aromatic rings, such as polyamides, will tolerate higher doses without deterioration. Because of the susceptibility of organic materials to radiation damage, great care is required in selecting and qualifying insulation materials and systems for use in these environments. The principle sources with which the designer is concerned are γ-rays and neutrons that interact with insulating materials to produce electrons that can be damaging.

The two molecular changes that may be produced by radiation in an organic insulation are cross-linking of molecular chains and bond scission or cutting of polymer chains. Cross-linking builds up the molecular structure, initially increasing tensile strength, but then reduces elongation and eventually results in the loss of impact strength. This changes rubbery or plastic material into hard, brittle solids. Scission breaks down molecular size, reduces tensile strength and usually adversely affects other properties. In outdoor applications, attention should be paid to the resistance of insulation to degradation by ultraviolet light that can also result in the embrittlement of the material.

2.6.6   *Synergism between ageing stresses*

Typically, the operating environment of most rotating machines results in two or more of these ageing mechanisms being present. Interactions between the stresses may be complex and result in unexpected consequences. The complexity introduced by a combination of stresses is one of the reasons that most insulation materials, systems and sometimes components are subjected to sequential application of individual stresses. Multifactor ageing of insulation materials, which tries to represent the operating conditions of an insulation system, is performed on sample insulation components in some cases; however, the results require significant analysis. An understanding of the synergistic effects between ageing stresses is central to the design of any condition-monitoring programme since it will affect the selection of the hardware and the interpretation of the data derived from these sensors and measuring instruments.

2.7   Insulation failure modes

2.7.1   *General*

From the preceding sections it can be seen that the means by which electrical machines fail depends not only on failure modes, which will be discussed in Chapter 3, but also on the type of machine and the environment in which it is working. However, it is possible to identify certain basic failure mechanisms that apply to all machines. We must then identify the early root causes of these faults because it is by detecting these root causes that monitoring becomes possible and beneficial.

Any fault involves a failure mechanism, progressing from the initial fault to the failure itself. The time taken for such a progression will vary, depending on a wide range of circumstances. What is important, however, is that all faults will have early indicators of their presence and it is here that monitoring must seek to look and act. Also, any fault is likely to have a number of possible causes and is likely to give rise to a number of early indications. A typical route to failure is shown in Figure 3.1.

In this and the following section we will give a description of the ailments with which electrical machines are afflicted. A detailed schedule of failure mechanisms for electrical machines is given in the Appendix where the failure modes and root causes for electrical machines are elaborated. By studying the Appendix the reader can identify what parameters should be worth monitoring in order to give an early indication of a particular fault. However, the general overview in the Appendix may fail to give the reader a clear image of the faults they may encounter on their machines. We will therefore describe in these sections a number of specific machine failure modes that draw out many of the factors that need to be considered in any monitoring system. Most of these incidents have been taken from the authors' experiences, so detailed references do not exist. However, various electrical machine users have tried to record their own failure statistics and typical results are summarised in References 4–9. Insulation faults present some of the most challenging problems for electrical machines, so this section will begin with the major insulation failure modes of electrical machines and then follow with other failure modes.

2.7.2   Stator winding insulation

2.7.2.1   General

Stator winding insulation is affected by all of the aforementioned stresses: thermal, electrical, environmental and mechanical; however, the extent to which these stresses in normal operation will cause problems in the short- or long-term will depend on factors such as the operating mode and type of ambient cooling conditions. For example, air-cooled machines tend to be subject to higher rates of thermal ageing compared to generators with direct liquid cooling of the stator winding. Further, generators with this type of cooling usually operate in a compressed hydrogen atmosphere thus eliminating oxidation.

2.7.2.2   Delamination and voids

Stator winding insulation is a laminated system consisting of numerous layers of mica-paper tape on a fibreglass backing material impregnated and consolidated with a synthetic resin, usually epoxy or polyester-based. The stator windings of rotating machines built prior to 1970 are likely to contain a bitumenous resin instead of epoxy or polyester. Further, these older insulation systems tend to employ large-flake mica rather than mica-paper. Consequently, the results derived from any condition-monitoring tool that is focused on stator winding insulation must be viewed in the context of the differences in these materials. While the older thermoplastic resins flow at relatively low temperatures, around 70 °C, causing resin migration leading to void formation and embrittlement of the main groundwall insulation, the high mica content provides significant resistance to partial discharge deterioration. Further, the propensity of bitumen-based insulation to soften at elevated temperatures permits the groundwall to conform to the walls of the stator core slot. The lack of this property in modern thermoset insulation systems based on synthetic resins resulted in the type of problems discussed next. Consequently, the use of age as a reliable index of machine health is not supported.

Voids or delaminations in the groundwall insulation of stator windings may result from the manufacturing process and/or operating stresses. The presence of voids in new stator windings, although not desirable and should be minimised, does not necessarily imply that the winding be rejected or that it is not fit for the design life intended. Application of various diagnostic tests, such as partial discharge and dielectric loss, as well as potentially destructive overvoltage tests, aid in the production of stator windings with minimal void content. During the life of the machine, these initial delaminations (or those initiated by thermal, mechanical or electrical stress) may result in the growth of voids that are prone to partial discharge. The probability that a void will be subject to partial discharge is governed by a number of factors such as void dimensions, electrical stress, pressure, temperature and the presence of initial electrons to cause discharge inception.

A partial discharge is so-called because it represents a gas breakdown between two dielectric surfaces or between a conductor and a dielectric surface. Details of the physics of the process can be found in Pedersen *et al.* [10]. Partial discharge, similar to any gas breakdown, results in the emission of heat, light and sound as well as the production of electrons and ionic species. Depending on the energy of the discharge, erosion of the void walls will result causing growth of the delamination and potentially failure in the long term. However, from a practical perspective stator winding failure caused by internal void discharge is very uncommon and is not considered by manufacturers and users of rotating machines to be a significant problem. There are two principle reasons that void discharge does not lead to rotating machine failure.

1. The presence of mica, a material that is extremely resistant to electrical discharge attack, results in the partial discharge erosion occurring only in the organic binding resin component. Consequently, the electrical breakdown path must follow a very circuitous route from the initiation site to the grounded core iron before failure can result. Typically, the initiation point is located at the edge of the copper conductor stack. Thus, the time-to-failure for such a process is very long, in the order of decades.

2. The widespread application of on-line partial discharge monitoring equipment as well as advances in the interpretation of the data produced by these tools. Use of partial discharge monitoring equipment has enabled machine users to better determine the condition of stator winding insulation in operating machines and to take corrective action at early stages. Although there is little maintenance available to remediate the presence of voids in the groundwall insulation, actions such as maintaining the integrity of the slot support system or reducing the operating stresses on the machine may arrest the process.

Thus, although void discharge has been and continues to be a subject of intense study by academic and other research organisations, failure of stator winding insulation from this mechanism is of low probability. The use of partial discharge measurements is of value, however, because partial discharge is a symptom of many other stator winding deterioration mechanisms that can cause failure. These mechanisms will be discussed below.

2.7.2.3   Slot discharge

Slot discharge or high-energy discharge is a very damaging deterioration mechanism found generally, but not exclusively, in air-cooled machines. Left undetected or uncorrected, failure from this type of mechanism can result in a relatively short period of time, two to three years. Slot discharge is the term used to describe a discharge occurring between the surface of the stator coil or bar and the grounded core iron. Generally, this mechanism results from a loss of good electrical contact between the insulated bar or coil surface and the stator core. Rotating machines rated above 3.3 kV employ a resistive coating applied to the slot portion of the stator coil or bar to promote good electrical contact with the core. Deterioration of this coating or loss of contact between bar surface and core iron can lead to conditions favourable for slot discharge.

The key to preventing slot discharge is to minimise differential movement between stator winding and core. Consequently, significant effort is expended by machine manufacturers and users to maintain the integrity of the slot support system; that is, the tightness of the stator wedges that restrain the winding in the slot. During the transition from thermoplastic to thermoset insulation systems, a significant increase in slot discharge problems on large air-cooled machines was observed by Lyles *et al.* [11,12]. The failure modes of large hydro generators, of the type shown in Figure 2.6, were described by Evans [13]. In these cases, the new hard insulation systems did not conform well to the stator slot unlike the older thermoplastic insulations. Consequently, significant vibration of the stator winding was encountered that led, during operation, to cyclical isolation of the stator bar or coil surface from the core iron. The resultant loss of electrical contact with the grounded core at one or more points along the coil or bar surface caused the surface potential on the bar to rise to line-to-ground potential. If this potential was sufficient to break down the gas gap between bar and core iron then a high energy discharge would result. Over a period of time these discharges would erode the resistive surface coating leaving isolated non-conducting areas that further exacerbated the problem since these locations would become capacitively charged, resulting in further discharge and erosion of the groundwall insulation itself.

A further mechanism leading to slot discharge in machines with apparently good bar-core iron contact resulted from abrasion of the resistive coating and the creation of isolated insulating patches on the bar or coil surface [14,15]. Wilson *et al.* identified the critical conditions leading to slot discharge for this case and also demonstrated analytically that the problem could also occur throughout the winding and not be limited to the high voltage coils or bars. This finding was borne out by experience in the field that found visual evidence of slot discharge damage at, or close to, the neutral point of the windings of large hydraulic generators as described by Lyles *et al.* [11,12].

Slot discharge can be identified using on-line partial discharge measurements. The partial discharge behaviour of slot discharge tends to be characterised by

- large magnitude pulses with a predominance of one polarity depending on how the measurement is accomplished,
- the position of the occurrence of the partial discharge pulses, which may vary over the power frequency cycle since the discharge is occurring between a capacitively charged surface and the core iron and these surfaces are, due to vibration, not at well-defined positions with respect to one another,
- rapid increase in partial discharge magnitudes over a relatively short period of time, such as doubling of pulse magnitudes over six months.

The production of ozone in open-ventilated air-cooled machines is both a symptom of slot discharge and ultimately a life-limiting fault in itself. Ozone is a strong oxidising agent and will have a negative effect on insulating and conducting materials. However, very often those machines that are producing excessive ozone are removed from service and rewound because the concentrations of ozone exceed the limits prescribed by health and safety legislation for workers.

2.7.2.4   Stator end windings

From the perspective of electrical insulation, the end windings of rotating machines are prone to environmental and mechanical rather than electrical ageing. Assuming that factory and/or acceptance overvoltage tests have demonstrated that the end-winding clearances have been properly designed and manufactured, the key problem areas for end windings are as follows.

- *Surface contamination*. The design and testing of the clearances in end windings are performed from the context of clean and dry insulating surfaces. However, air-cooled machines, especially motors, often operate in ambient conditions containing high levels of conducting contaminants such as salt, cement dust, lubricating oil and brake dust. Over time, and in the presence of moisture, these conductive layers distort the electrical field in the end winding and lead to surface tracking. This deterioration mechanism if not corrected may lead to serious problems including phase-to-phase failure of the machine. Partial discharge measurements can be used to detect the presence of contaminated end windings.
- *Internal voids*. Until relatively recently, the end-winding portion of stator bars and especially stator coils tended to be taped manually rather than by machine, as is the case in the slot cell. Consequently, the insulation of the end-winding region is prone to contain voids or delaminations. The presence of such faults is not as critical as for the slot cell of the winding because the electrical stress in this area is relatively low.
- *Mechanical aspects*. The end-winding structure of a rotating machine, especially for large turbine generators, represents several challenges for designers. Apart from electrical clearance, the major issues are minimising vibration and differential movement. Modern synthetic insulation systems are quite rigid and thus prone to cracking either due to long-term cyclic fatigue in normal operation or transient events causing end-winding distortion such as a close-in transformer fault. Unfortunately, while the behaviour of insulating materials in this region of the machine is considered of significant interest, relatively little work has been done to develop a similar understanding of the creep and fatigue properties that exist for metals.

2.7.2.5 End-winding stress grading

End-winding stress grading systems tend to be found on machines rated 6.6 kV and above. Their role is to reduce the electric field at the surface of the stator bar or coil where it exits the core slot. Without application of such a material, there is a high probability of surface discharge or even flashover from the capacitively charged insulating surface back to the core iron. The functionality of the stress grading system may be compromised by surface contamination or improper design and processing of the interface between the end-winding grading and slot stress control systems. In the latter case, a combination of electrical and thermal effects can cause erosion of the junction between the two systems, potentially resulting in an increase in surface discharge activity. While this fault can appear serious the discharges tend to be transverse to the surface resulting in a low probability of failure due to this mechanism. Empirical evidence indicates that when the junction between end-winding and slot stress control systems has been completely disrupted, and the electrical contact lost, the surface discharges cease.

2.7.2.6   Stator winding inter-turn faults

A common failure mechanism on machines employing multi-turn stator coils is breakdown of the turn insulation. The resultant short circuit between the copper turns causes a significant circulating current to flow in the coil leading to rapid deterioration and failure. Turn failures tend to be very destructive, and involve burning of the insulation and localised melting of the copper conductors. Often, failures resulting from breakdown of the inter-turn insulation are inferred from the location of the puncture, typically at or near the core exit, and the electrical position in the winding, typically the first or second coil from the line end.

During the 1980s, an Electric Power Research Institute study investigated the reasons for an apparent increase in the number of turn failures experienced by North American utilities. Initially, the increase in breakdowns was attributed to the increase in the use of vacuum circuit breakers as replacements for conventional magnetic air circuit breakers. The steep-fronted, high-magnitude transients generated during operation of the vacuum circuit breaker was considered to exceed the capability of the turn insulation. However, the Electric Power Research Institute study concluded that the principal reason for the perceived increase in turn insulation failures was inadequate turn insulation. Those machines that incorporated dedicated turn insulation consisting of either mica-paper tape or double Daglas were found to have superior capability of withstanding the steep-fronted surges imposed by vacuum circuit breakers as well as being able to withstand the rigours of steady-state operation at power frequency. The use of dedicated turn insulation provides abrasion resistance in the event of differential movement between turns due to thermal or thermomechanical ageing.

Early detection of the onset of conditions likely to lead to breakdown of inter-turn insulation may be possible using partial discharge methods and/or techniques based on axial leakage flux detection, investigated by Penman *et al.* [16,17]. With regard to partial discharge-based methods, a significant practical problem is that although the deterioration of the inter-turn insulation can, in principle, be detected, attributing the detected signals to this deterioration mechanism in the presence of partial discharge pulses originating from other faults is practically impossible.

2.7.2.7   Repetitive transients

Variable speed drives based on electronic inverters are widely applied on low-voltage (600 V and below) motors. Insulation problems have been experienced on these largely random wound machines resulting in failures. The bulk of the failures have been attributed to electrical discharges in the end windings of these machines. The steep-fronted surges generated by the insulated gate bipolar transistor (IGBT) devices employed in these drives may cause voltage doubling resulting in transient voltages that exceed the electric field required to cause breakdown in the air around the end windings. The insulation systems employed in random wound machines, typically polyester-based enamels and Daglas, have poor discharge resistance and are thus prone to failure in the presence of such discharges. Steps to mitigate this problem include: better design of the winding, to minimise areas of high electrical stress; consideration of the length of cables used to connect to the motors, to reduce the probability of voltage doubling; the use of multi-converter topologies; and the use of metal-oxide loaded enamels to grade the winding electric stress. High-voltage motors, employing form wound stator windings, are less prone to the deterioration mechanism described on p. 42 although some consideration is being given to the behaviour of the resistive slot stress control material under repetitive transient conditions.

2.7.3   Stator winding faults

2.7.3.1   Stator winding faults (all machines)

It is clear that the insulation system is potentially one of the intrinsically weakest components of an electrical machine, both mechanically and electrically and in the earliest days of machine construction insulation faults were excessively frequent, as described by Walker [18].

However, as described on p. 20, modern techniques of winding manufacture, using thermosetting resin or vacuum pressure impregnated insulation, provide systems that are mechanically tough and electrically sound. Nonetheless, modern machine use drives insulation systems to their thermal, mechanical and electrical limits. However, there are very few incidences of failures on machines due simply to ageing of insulation. A very thorough review of root causes of failures and failure modes in insulation systems and conducting components is given in Stone *et al.* [19]. An example of a generator winding failure is shown in Figure 2.11 and the final puncture of the insulation system was the result of accelerated ageing due to an over-temperature

Main wall insulation puncture

Figure 2.11   Stator fault in a 15 kV generator winding. View shows the main-wall insulation punctured in the top right as a result of core overheating.

in the adjacent generator core. An exception to this may be on air-cooled, highly rated hydro-generators where epoxy mica insulation has failed due to erosion of the insulation in the slot portion as a result of discharge activity, primarily because of the rigidity of epoxy mica systems, the large forces on such large conductor bars, the high dielectric stress on the windings and the fact that these machines are air-cooled. However, failures due to isolated insulation faults do occur, due to manufacturing faults such as voids or foreign bodies embedded in the main wall insulation, or penetration of the insulation by foreign material such as oil or metal from elsewhere in the machine.

Whether insulation failure occurs due to ageing or the action of an isolated fault, the indications are similar in that there will be an increase in discharge activity in the machine culminating in an insulation puncture like that shown in Figure 2.11. Faults can also occur in low voltage windings, which are generally dipped in varnish and do not experience high-voltage discharge. Figure 2.12 shows a typical failure in a round wire winding that is varnish impregnated showing the effect of moisture contamination and tracking that has eaten away part of the winding slot liner causing a fault to earth.

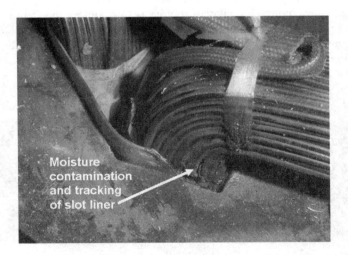

Figure 2.12   Stator fault on a 200 V exciter winding. View shows the effect of moisture contamination and tracking.

2.7.3.2   Winding conductor faults (generators)

These faults are again generally confined to large generators where the current densities are such that the stator winding is electrically, thermally and mechanically highly stressed. It is normal to subdivide the conductor into a large number of subconductors and to insulate and transpose them to minimise winding losses. In the most modern machines the transposition is distributed throughout the conductor length employing the Roebel technique. This gives a uniform current distribution and minimises the voltages between subconductors. Older machines, however, have the transpositions made in the end winding at the knuckle joint and quite large voltages, up to 50 V root mean square (rms), can exist between subconductors. If severe mechanical movement of the winding occurs during operation and the subconductor insulation fails then subconductors can short together causing arcing. A number of older machines have failed in the UK and US due to this arcing, which in the worst cases has eroded and melted other subconductors, pyrolysing the main wall insulation of the conductor bar. If this happens in the slot portion or near to other earthed metalwork, then an earth fault can occur but if not, it is possible for the debris ejected from the burnt area to produce a conducting path between electrical phases of the winding leading to the more serious phase-to-phase fault, as shown in Figure 2.13. Subconductor arcing has also been initiated by fatigue failure of subconductors, where the conductor bar emerges from the core or enters a water box, due to excessive and winding vibration. The consequences of such a failure are very similar to those of a subconductor short. When this arcing takes place in a machine with hollow water-cooled subconductors, perforation of a subconductor occurs and, since they are usually arranged so that there is an excess of cooling gas pressure over the water, this leads to a leakage of gas into the water-cooling system. The early indicators of such faults are arcing activity within

Figure 2.13   *Example of the effects of a stator subconductor fault in a 588 MVA, 23 kV, 50 Hz turbine generator winding. View shows stator conductor bars ejected into the generator air-gap and seriously disrupting the end winding. [Taken from Tavner et al. [7]]*

the winding and the pyrolysing of insulation. The fact that burning takes place deep within a conductor bar usually means that, initially at least, only small quantities of particulate and gaseous matter produced by the burning are released into the cooling gas circuit. Where water-cooled subconductors are present the fault causes a leakage of gas into the winding cooling system.

2.7.3.3   Stator end-winding faults (all machines)

A considerable amount of effort over the whole range of machines has gone into the design of end-winding structures. The first objective is to restrain the end winding against the large forces on the winding during transient loading or faults and the second is to cushion the conductor bars against the smaller forces during steady, continuous running. End-winding movements are larger on the older, less rigid, bituminous mica insulation system but then that system, because of its softness, was more able to withstand the steady fretting action of normal running than the hard epoxy-mica systems. In either case end-winding movements in normal operation are quite significant and can be as much as a few millimetres on a large turbine generator.

Faults occur in the end winding when the bracing structure slackens, due to torsional and lateral vibration, either as a result of a succession of unusual overloads or because of an extended period of continuous running. In some cases end-winding insulation becomes cracked, fretted or worn away. In the limit a full line to line fault can lead to a catastrophic failure of the end winding as shown in Figure 2.13. On the largest machines fatigue failure of conductors can occur when the winding becomes slack enough to permit a significant amount of conductor movement during normal operation or during the much larger forces of starting or reswitching. In addition, seismic excitation of the relatively weak end-winding structure can occur in applications with inherent vibration where the machine is mounted on a flexible foundation and this has been a major cause of end-winding failures in the offshore industry.

Foreign bodies inside a machine such as steel washers, nuts or small portions of insulation, get thrown around by the rotor. Damage is caused by these objects, usually in the stator end-winding region, where the insulation is damaged by impact or eroded by debris worming into the insulation under the action of electromagnetic forces. The early indications of problems are an increase in end-winding vibration and the possibility of electrical discharge activity to nearby earth planes.

2.7.3.4   Stator winding coolant system faults

Many large machines employ direct or indirect liquid cooling of the stator windings to permit operation at higher energy densities. In the vast majority of cases, the coolant medium is water; however, a small number of designs use oil. Consequently, there is a risk that the coolant path will deteriorate resulting in the release of water, of varying levels of purity, onto insulating or high voltage components resulting in deterioration and potentially failure of the machine.

Large steam turbine generators employ direct liquid cooling of the stator windings in that demineralised water flows through hollow subconductors. Most designs, whether single- or two-pass, use a manifold at one or both ends of the winding that connects to the individual stator bars using insulated (Teflon) hoses (see Figure 2.5(a)). Catastrophic failure of the hose or the connection of the hose to the bar or manifold will result in water impinging on the stator winding. Depending on the condition of the water, such an event may not cause an immediate stator earth ground alarm or fault; however, the coolant loop pressure and stator winding temperature will be affected. Smaller leaks would likely result in gas-in-coolant alarms and potentially gas-locking.

Some types of turbine generators employ a plenum that connects a group of stator bars directly to the water circuit without the use of hoses. The plenum consists of cast insulating epoxy components, mechanical failure of which will cause loss of the coolant circuit and water impingement on the stator winding. Other sources of potential water leaks onto the stator winding include faults in the hydrogen coolers. In this case, the water is likely to have a relatively high conductivity and would have a high probability of causing a stator ground fault alarm and ultimately fault.

Water leaks into the bulk of the stator winding insulation can also occur. This water ingress has been observed on liquid-cooled stator windings using the clip design where the connection of the stator bar to the coolant system is affected. Pinhole leaks appear after a few years of service and moisture propagates along the brazed joint until contact with the strand insulation. The moisture further wicks into the insulation degrading the dielectric properties. Two principle root causes for moisture ingress into stator winding insulation are

- crevice corrosion, resulting from chemical reaction between the coolant water and the braze material,
- deficiency in the quantity of brazing material. The initial brazed joint is tight, but over time vibration and thermal stresses cause voids in the braze to connect creating a channel for the water.

In principle, the presence of a leakage channel in the brazed region should result in a gas-in-coolant alarm. However, the insulation covering the water box is very tight and hydrogen does not migrate easily through it. Therefore, water box leaks detected in this way are rare.

Water ingress in the insulation wets the material to up to ten times the initial moisture level. Another consequence of water ingress in the groundwall insulation is a drop in insulation resistance. This resistance decay can be quite slow. It is considered that it takes about five years for water to migrate and cause a significant insulation drop. This is the main mechanism for stator winding failure due to water box leaks. Several stator earth fault failures are due to water ingress and thus to water box leaks. So far, water box leak detection methods have been based on detecting one of the consequences of the water ingress. However, no reliable on-line method has been developed.

In addition to problems caused by coolant leaks, serious damage can be caused by blocking of the coolant flow in stator windings. Depending on the location and number of plugged subconductors catastrophic failure due to overheating can occur rapidly. Other than blockages caused by foreign material entering the coolant loop and gas locking, flow restrictions can result from the deposition of copper oxide deposits. These deposits may result from poor control of stator water chemistry.

Plugging or flow restriction of the coolant medium in the hollow subconductors of liquid-cooled stator windings can have serious if not catastrophic implications for the reliable operation of large steam turbine generators. Significant efforts have been and continue to be expended on understanding the fundamental mechanisms resulting in coolant flow restriction. Apart from loose debris in the coolant circuit, the predominant cause of blockage is corrosion of the copper conductors resulting in the deposition of copper oxides that tend to clog the small-aperture hollow subconductors. Consequently, much of the work in this field has focused on the chemistry of the coolant water.

Unfortunately, for some utilities plugged subconductors are already a reality, resulting in the derating of units and/or the implementation of complex flushing and cleaning procedures. Again, many organisations are pursuing work to understand and optimise these remedial actions. Consequently, significant work is underway, or has already been performed, to understand the plugging mechanism and the means to prevent or remedy it. These efforts are necessary to enable utilities to maximise reliability and availability of their generators. However, there is a third aspect to this problem; namely, for those utilities experiencing overheating due to plugged subconductors, how best to manage the situation so as to minimise revenue loss while ensuring that the stator winding suffers no undue loss of life.

Early detection of stator bar overheating due to plugging is complicated because of, in many cases, the dearth of sufficient temperature and flow data. Even where significant quantities of temperature data exist, such as on hose-type machines with thermocouples on the outlets, the location of the temperature sensor with respect to the hot spot increases the difficulty of successful early plugging detection. Some organisations have recognised this problem and have attempted to address it by detailed thermohydraulic modelling of the stator winding and/or the implementation of post-processing of the temperature. Notwithstanding these efforts, operators of large steam turbine generators presently lack tools to objectively assess the risks associated with operating machines with known or possible subconductor plugging.

2.7.4   Rotor winding faults

2.7.4.1   General

The rotor windings of generators are insulated with epoxy-glass laminates or polyester-based materials (Nomex). Squirrel cage induction motor rotor windings consist of lightly insulated copper bars driven into the slots in the case of a laminated rotor or aluminium bars cast directly into the rotor. The principal stresses of concern on rotor windings are thermal and mechanical.

2.7.4.2   Induction motor rotor faults

Faults on the rotor windings of induction motors have not been easy to detect because there is not necessarily an electrical connection to the winding and it is difficult to measure the low-frequency currents induced in the rotor winding. Although the rotor winding of a squirrel cage induction motor is exceptionally rugged, faults do occur particularly on the larger machines when they are subjected to arduous thermal and starting duty, which causes high temperatures in the rotor and high centrifugal loadings on the end rings of the cage (see Figure 2.14).

Broken bars – note: bars have lifted from slots

Ventilating ducts

End ring

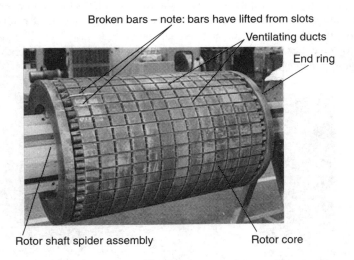

Rotor shaft spider assembly

Rotor core

Figure 2.14   Rotor cage fault in a 1.5 MW squirrel cage induction motor [Taken from Thomson and Fenger [20]. © IEEE (2001)]

Faults may occur during manufacture, through defective casting in the case of die cast rotors, or poor jointing in the case of brazed or welded end rings. Such a fault results in a high resistance, which will cause overheating, and at high temperatures the strength of the cage will be impaired. Cracking may then occur in the rotor bar and indeed usually takes place at the cage end rings where the bars are unsupported by the rotor core. Similar faults can occur because of differential movement of the cage in the rotor slots, because of a succession of periods of high temperature running and shutdowns. This can lead to distortion and ultimately cracking of the end rings and the associated bars. It should be remembered that the bars must provide the braking and accelerating forces on the end ring when the motor changes speed. If the motor speed fluctuates, because of changing load or as part of the normal duty cycle, then high-cycle fatigue failures can occur at the joints between bars and ring. If the motor is repeatedly started then the exceptional starting forces may lead to low-cycle fatigue failure of the winding component. The early indications of these faults are pulsations in the speed, supply current and stray leakage flux of the machine.

The rotor windings of wound rotor induction motors are of rather similar design to the stator winding of the motor except that the end windings must be restrained against centrifugal forces by steel wire or woven glass fibre banding rings. Damage to the windings usually occurs in the end region due to centrifugal forces on the crossovers and connections of the winding causing shorts between turns. These faults are similar to the problems experienced on the rotor windings of high-speed turbine-type generators. An additional difficulty encountered with the wound rotor machine is that of ensuring balance between the phases of the external resistors connected to the winding via the slip rings. These resistors are usually water-based and can become unbalanced with time, causing the currents flowing in the rotor windings to become unbalanced and overheating to occur. This can lead to the rapid degradation of rotor winding insulation and ultimately failure such as that shown in Figure 2.15. The problem is difficult to detect because the rotor currents are at a very low slip frequency and one must be able to detect relatively small changes in these currents.

Figure 2.15   *Rotor winding fault in a 3 MW slip ring induction motor. View shows arcing damage between the phases in the winding end region. [Taken from Tavner et al. [7]]*

2.7.4.3   Turbine generator rotor winding faults

The rotor winding insulation and bracing system in turbine generators must be designed to withstand the extremely high centrifugal forces present during operation. An inter-turn short-circuit can occur between rotor turns due to puncture or cracking of the turn insulation. The current that subsequently flows between turns creates localised heating and the probability of further shorted turns. This disturbance of the current flow will cause an asymmetry in the flux in the machine, which causes unbalanced magnetic pull (UMP) leading to vibration of the rotor and this can be compounded when the asymmetric heating leads to thermal bending of the rotor.

Two- and four-pole totally enclosed machines are particularly prone to these problems especially if they have a short airgap that will have a reduced coolant gas flow and therefore rotor temperatures can be higher. Differential movement between turns and fretting action in the rotor winding can produce copper dust, which may also increase the probability of turn shorts. This phenomenon occurs because of the cyclic movement that a large winding experiences relative to the rotor. The movement is partly caused by self-weight bending of the rotor and thermal cycling. This problem can also be exacerbated by long periods of barring or turning gear. During this type of very low speed operation, there are little or no centrifugal forces that act to lock the key internal rotor components in place and hence minimise differential movement.

If an insulation fault occurs between the winding and rotor body then a ground fault current flows that can be detected by an earth leakage relay. The ground fault current is limited such that a single ground fault is not serious but a second ground fault can result in very large circulating currents. A second ground fault can cause an arc to be struck at the fault location not only causing winding damage but also severe damage to the rotor forging. The early indications of these faults are distortion of the airgap flux and associated stray leakage flux around the machine, and an increase in bearing vibration. Therefore the early indications of these faults are usually excessive transverse bearing vibrations, although attention has also focussed on measuring the torsional oscillations of the shaft itself.

Figure 2.16   Failure of a DC armature winding close to the commutator

2.7.4.4   Rotor winding faults (DC machines)

DC machines encounter particular difficulties in their armature, where the AC winding is directly connected to the commutator and electrical action between brushes and commutator segments can lead to high temperatures and the production of carbon dust. A common problem with DC armature windings is failure between conductors close to the commutator connection (Figure 2.16 shows just such a failure) possibly aggravated by the accumulation of carbon dust under the banding that provides the centripetal force to retain the winding in place close to the commutator connections.

2.8   Other failure modes

2.8.1   Stator core faults

A core fault is a rare event, which usually only occurs in the largest turbine-driven generators where the laminated steel cores are sufficiently massive, and carry a sufficiently high magnetic and electric loading (see Table 2.4). They can occur anywhere in the laminated steel core of large rotating electrical machines but are more common in the stator core [2]. A core fault initiates when core plates are electrically connected together, either because of an insulation failure between them, or as a result of physically imposed short circuits due to various causes including foreign bodies. These connections circulate additional currents in the core, coupling the main flux, leading to additional losses and heating, further weakening core plate insulation and expanding the fault. The fault is then at a zone in the core that is overheated, buckled and electrically interconnected. This can expand, melting material, leading to the catastrophic runaway of the fault and creating a cavity in the core, continuing until the main conductor insulation is damaged and the machine is disconnected for an earth fault. Core faults generally occur some distance from the main conductors and final disconnection frequently occurs after irreparable damage has been done to the core and conductors. A number of large generators worldwide have experienced this problem (see Figure 2.17) mainly as a result of damage to the their stator cores during manufacture or rotor insertion, when laminations became shorted together, and the application to the core of transient high flux conditions.

2.8.2   Connection faults (high-voltage motors and generators)

Faults in the connections of electrical machines are very unusual in the lower voltages (<1000 V rms) but become more common at higher voltages where the dielectric stress on the bushings and forces on the conductors increase.

Insulation failures also occur from time to time on the bushings that carry the electrical connections through the machine enclosure. On a turbine-type machine these are mounted in the pressure-tight casing and therefore have to withstand the operating pressure of the machine. Failure of a bushing can occur either due to mechanical stresses or vibration on the conductor passing through the bushing, causing it to crack, or due to debris being deposited on the exposed surfaces allowing it to track electrically, as described in Section 2.6.3.3. Again the early indication is an increase in discharge activity.

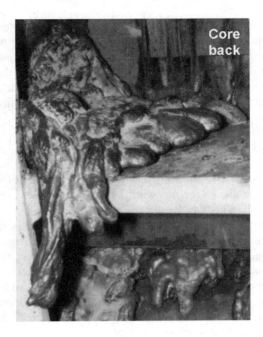

Figure 2.17   Core fault in the stator core of a 660 MW turbine generator [Taken from Tavner and Anderson [2]]

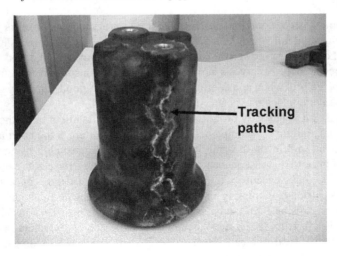

Figure 2.18   Tracking damage on a cast resin bushing similar to that used for the connections of high-voltage motors and generators

Figure 2.18 shows damage to a high-voltage switchgear bushing, similar to that used in high-voltage machines, where moisture and contamination have allowed a conductive film to form on the surface of the bushing and severe surface tracking has taken place. The high dielectric stress on the surface of the bushing has allowed tracking to take place over an extended period of time, probably many months and possibly even more than a year. This kind of damage tends to occur where bushings are exposed to the environment outside the machine. It is rare inside the machine unless there is a cooling-liquid leak combined with solid contamination circulating in the inner cooling circuit. This kind of damage can be diagnosed by physical inspection, partial discharge measurements or thermography.

2.8.3   Water coolant faults (all machines)

In machines with water-cooled heat exchangers it is possible for coolant blockage or failure to occur, either in the pipework leading to the heat exchanger or in the heat exchanger itself. This can be the result of pipework debris being circulated in the water system (although this should be removed by filters fitted close to the inlet of the machine) or due to negligence when cooling systems or machines are maintained. The normal vibration of a machine in service can excite resonances in an improperly designed cooling pipework system and this can cause fatigue failure of a pipe and loss of coolant. The early indications of these kinds of fault are indicated by high conductor or cooling water temperatures, leakage from the water system and pyrolysing insulation eventually leading to damaging discharge activity that will be electrically detectable at the machine terminals.

2.8.4   Bearing faults

Rolling element, sleeve and pad bearings are used in rotating machines as guide and thrust bearings and fail when the load upon the bearing is excessive or its lubrication fails. The choice of bearing and the lubrication used depends upon the load borne and the shaft speed. Table 2.6 gives a summary of the conditions involved in the choice. These can be helpful when considering fault conditions. A good summary of bearing selection is given in Neale [21] and shown in Figure 2.19.

Bearing failure is usually progressive but ultimately its effect upon the machine is catastrophic. Failure is accompanied by a rising temperature at the bearing surface, in the lubricant and in the bearing housing, which are detectable by temperature sensors. An important consequence of bearing deterioration for electrical machines is the rotor becomes eccentric in the stator bore causing a degree of static and/or dynamic eccentricity, disrupting the fine balance between the magnetic forces of adjacent poles, causing UMP and placing more load on the bearing. This also causes an increase in vibration as the shaft dynamics are affected by the altered airgap and bearing stiffness. Bearings can also be damaged by the flow of shaft currents as described in the following section.

2.8.5   Shaft voltages

Electrical machines can generate voltages in their shafts due to an unbalance in the magnetic circuit of the machine [22]. The shaft voltage can circulate current through

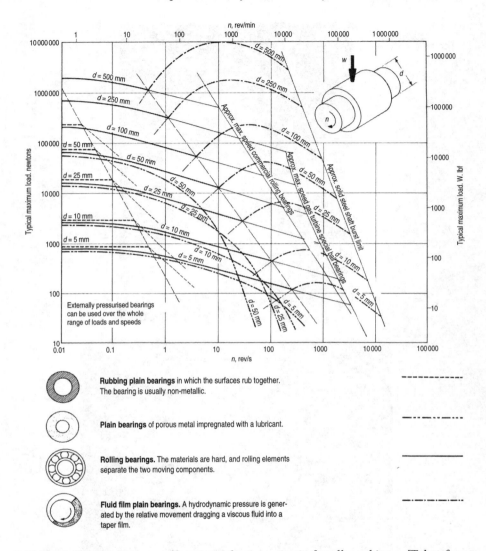

Figure 2.19   A summary of bearing selection criteria for all machinery [Taken from Neale [21]]

the bearings of the machine and this leads to bearing damage. Voltages on small machines such as those shown in Figure 2.9 can be of the order of 500 mV rms, but in larger machines such as those in Figures 2.1 and 2.2 they can be 5–10 V rms.

There is a considerable literature about shaft voltages and manufacturers design machines to minimise their effects by insulating one bearing of the machine and providing a shaft ground to allow the shaft voltage to be monitored. The damage that occurs to bearings depends upon the type of bearing. In general a rolling element

(a)

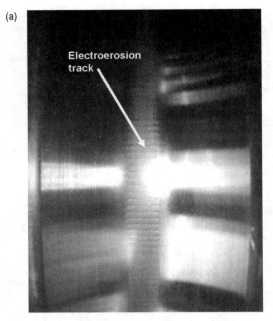

(b)

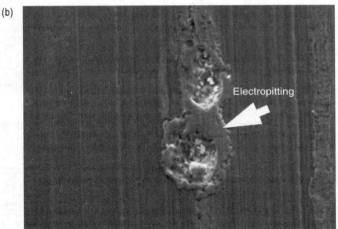

Figure 2.20   (a) Electroerosion damage in the race of a rolling element bearing due to the flow of shaft currents in a generator. (b) Electropitting damage to a bearing sleeve due to the flow of shaft currents in a generator.

bearing is more susceptible to the effects of shaft voltages, exhibiting electro-erosion (see Figure 2.20(a)). Sleeve bearings have a higher impedance but also can be damaged by the pitting action consequent upon the flow of shaft current (see Figure 2.20(b)).

2.9 Conclusion

This chapter has shown the common structure present in electrical machines regardless of their size. The construction of electrical machines has also been demonstrated with an indication of the effect of operational service on failure modes. The importance of the insulation system of the machine has also been considered and discussed. Finally, typical failures in service have been shown to identify the most common failure modes that an engineer will encounter, starting with those originating in the insulation system but then expanding to consider other sources. The failure modes demonstrate how faults may be detected in their early stages by monitoring appropriate parameters. In the next chapter we will describe reliability analysis, which will connect root causes to failure modes and shows how condition monitoring can be directed to address particular components in the electrical machine.

2.10 References

1. IEC 60085. Method for determining the thermal classification of electrical insulation. BS 2757:1986. British Standards Institute, London, 1984.
2. Tavner P.J. and Anderson A.F. Core faults in large generators. *IEE Proceedings Electric Power Applications* 2005; **152**: 1427–39.
3. Bone J.C.H. and Schwarz K.K. Large AC motors. *IEE Proceedings* 1973; **120**: 1111–32.
4. Dickinson W.H. IEEE Reliability Working Group. Report on reliability of electric plant, parts I, II and III. *IEEE Transactions on Industry Applications* 1974; **IA-10**: 201–52.
5. Barker B. and Hodge J.M. *A Decade of Experience with Generator and Large Motor Reliability*. Paris: CIGRE; 1982.
6. O'Donnell P. IEEE Reliability Working Group. Report of large motor reliability survey of industrial and commercial installations, parts I, II and III. *IEEE Transactions on Industry Applications* 1985; **IA-21**: 853–72.
7. Tavner P.J., Gaydon B.G. and Ward D.M. Monitoring generators and large motors. *IEE Proceedings, Part B, Electric Power Applications* 1986; **133**: 169–80.
8. Thorsen O.V. and Dalva M. A survey of faults on induction motors in offshore oil industry, petrochemical industry, gas terminals, and oil refineries. *IEEE Transactions on Industry Applications* 1995; **IA-31**: 1186–96.
9. Albrecht P.F., Appiarus J.C., McCoy R.M., Owen E.L. and Sharma D.K. Assessment of the reliability of motors in utility applications – updated. *IEEE Transactions on Energy Conversion* 1986; **EC-1**: 39–46.
10. Pedersen A., Crichton G.C. and McAllister I.W. The theory and measurement of partial discharge transients. *IEEE Transactions on Electrical Insulation* 1991; **EI-26**: 487–97.
11. Lyles J.F., Goodeve T.E. and Sedding H. Parameters required to maximize a thermoset hydro-generator stator winding life. Part I – design, manufacture, installation. *IEEE Transactions on Energy Conversion* 1994; **EC-9**: 620–27.

12.   Lyles J.F., Goodeve T.E. and Sedding H. Parameters required to maximize a thermoset hydro-generator stator winding life. Part II – monitoring, maintenance. *IEEE Transactions on Energy Conversion* 1994; **EC-9**: 628–35.
13.   Evans D.L. IEEE Reliability Working Group. Report on problems with hydro-generator thermoset stator windings, Part I, II and III. *IEEE Transactions on Power Apparatus and Systems* 1981; **PAS-100**: 3284–303.
14.   Jackson R.J. and Wilson A. Slot-discharge activity in air-cooled motors and generators. *IEE Proceedings, Part B, Electric Power Applications* 1982; **129**: 159–67.
15.   Wilson A. Slot discharge damage in air cooled stator windings. *IEE Proceedings, Part A, Science Measurement and Technology* 1991; **138**: 153–60.
16.   Penman J., Sedding H.G., Lloyd B.A. and Fink W.T. Detection and location of interturn short circuits in the stator windings of operating motors. *IEEE Transactions on Energy Conversion* 1994; **EC-9**: 652–58.
17.   Penman J., Hadwick G. and Stronach A.F. Protection strategy against faults in electrical machines. Presented at the International Conference on Developments in Power System Protection, London, 1980.
18.   Walker M. The Diagnosing of Trouble in Electrical Machines. London: Library Press; 1924.
19.   Stone G.C., Boulter E.A., Culbert I. and Dhirani H. Electrical Insulation for Rotating Machines, Design, Evaluation, Aging, Testing, and Repair. New York: Wiley–IEEE Press; 2004.
20.   Thomson W.T. and Fenger M. Current signature analysis to detect induction motor faults. *IEEE Industry Applications Magazine* 2001; **7**: 26–34.
21.   Neale M.J. *Tribology Handbook*. London: Butterworth; 1973.
22.   Tavner P.J. Permeabilitatsschwankungen auf Grund von Walzeffekten in Kern-platten ohne Valenzrichtung. *Elektrotechnik und Maschinenbau* 1980; **97**: 383–86.

Chapter 3
Reliability of machines and typical failure rates

3.1 Introduction

Chapter 2 described the construction and operation of electrical machines and the way they fail in service. We now need to move towards identifying the condition monitoring that needs to be applied to detect those faults.

Condition monitoring has been seen primarily as the province of those who analyse fault signals and interpret results. However, interpretation must be informed by causality and the authors' view is that one of the important changes to occur in condition monitoring over the last 30 years is ensuring that it is firmly directed towards the root causes of machine failure. In so doing, successful condition monitoring directly addresses machine unreliability or, more importantly for the operator, lack of availability. This chapter sets out the issues concerned with this causality.

3.2 Definition of terms

In considering the progression to failure of machinery it is necessary to define a number of useful terms.

- *Failure* is when the electrical machine fails to perform its energy conversion function. Failure is complete and does not imply partial functionality.
- *Failure mode* is the manner in which final failure occurred. For example
  - insulation failure to ground,
  - structural failure of a shaft.
- *Root cause* is the manner in which the failure mode was initiated. For example for the two cases above
  - overheating of the insulation could be the root cause leading to the failure mode of insulation failure to ground,
    or
  - excessive shock torque being applied to the shaft could be the root cause leading to the failure mode of structural failure of the shaft.
- *Failure mechanism* is the physical manner in which a failure process progresses from the root cause to the failure mode. For example in the two cases above
  - overheating causes degradation of the insulation material leading to reduced voltage withstand capability,
    or

- excessive shock torque on the shaft causing yield and an increase in stress in the remaining parts of the shaft leading to a progressive and ultimately catastrophic yield of the component.
- *Root cause analysis* (RCA) is the analysis following a failure of the failure modes and underlying root causes.
- *Duration of the failure sequence* is considered in this book to be the time from root cause to failure mode. This may be a period of seconds, minutes, days, months or weeks and can depend on
  - the failure mode itself,
  - the operating conditions of the machine,
  - the ambient conditions.
- *Reliability* (R) is the probability that a machine can operate without failure for a time t and in this book is generally quoted as a failure rate. A high reliability machine has few failures, a high mean time between failure, a high percentage reliability and a low failure rate. The reliability as a function $R(t)$ is sometimes known as the survivor function because it indicates what proportion of the starting population survives at a particular time t. A word definition of reliability would be 'the probability that a system will operate to an agreed level of performance for a specified period, subject to specified environmental conditions'.
- *Availability* (A) is the probability that a machine will be available to operate for a time t and in this book is generally quoted as a percentage. A high availability machine has only short periods of time shut down due to failure or maintenance.
- *Failure rate* ($\lambda(t)$) is the rate at which failures occur in a machine and in this book is generally quoted as failures/machine/year. Failure rate is sometimes considered to be constant throughout machine life but in fact it varies according to the operating condition and age of a machine. The failure rate as a function of time is sometimes called the *hazard function*. It is the objective of the maintenance engineer of the machine to keep the failure rate low, constant and predictable.
- *Time to failure* (TTF) is the time measured from the instant of installation of the machine to the instant of failure. The *mean time to failure* (MTTF) is the expected value of that and successive TTFs. MTTF does not include the time to repair as a result of a failure. The MTTF is usually given in hours.
- *Time to repair* (TTR) is the time measured from the instant of first failure to the instant when the machine is available for operation again. The *mean time to repair* (MTTR) is the average of that and successive TTRs and can be averaged over a number of machines in a population. The MTTR is usually given in hours.
- Under the hypothesis of minimal repair, that is repair that brings the unit back to the condition before failure, *time between failure* (TBF) is the time measured from the instant of installation of the machine to the instant after the first failure when the machine is available for operation again. The *mean time between failure* (MTBF) is the average of that and successive TBFs and can be averaged over a number of machines in a population. MTBF is the sum of the MTTF and MTTR. The MTBF or θ is usually given in hours.
- *Failure mode and effects analysis* (FMEA) is a subjective analysis tool, defined by US standards [1] that uses a qualitative approach to identify potential failure

modes, their root causes and the associated risks in the design, manufacture or operation of a machine.

3.3   Failure sequence and effect on monitoring

The failure sequence from operation to failure for a specific failure mode in a typical subassembly or component, in this case the main shaft of the machine, can be drawn vertically as shown in Figure 3.1 and contains the sequence root cause–failure mode– failure.

The duration of the failure sequence depends on the failure mode, the operating condition of the machine and the ambient condition in which it is operating.

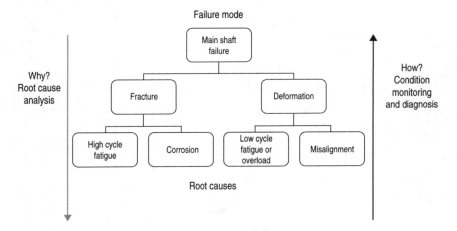

Figure 3.1   *Cause-and-effect diagram showing the relationship between the failure sequence and root cause analysis for an example failure, the fracture of a main shaft initiated by stress corrosion cracking*

Figure 3.2 demonstrates this process for a failure mode described by a normal distribution. Figure 3.2(a) shows the progression from reliable to non-reliable opera- tion rapidly at the 50 per cent life point for a fast fault. The probability of failure rises sharply close to this point, the area under the curve being equal to 1 because in the life of the machine there is 100 per cent probability of failure. Figure 3.2(b) repeats the sequence for a medium speed fault. Finally Figure 3.2(c) repeats the sequence for a slow speed fault.

This process is at the heart of condition monitoring. If a failure sequence is very rapid; for example, a few seconds like the fast fault in Figure 3.2(a), then effective condition monitoring is impossible. This is the situation for most electrical failure modes which actuate the electrical protection, where the period of action of the final failure mode may be only a few seconds or even only a few cycles of the mains. However, if the failure sequence is days, weeks or months, like the slow fault in Figure 3.2(c), then condition monitoring has the potential to provide early warning

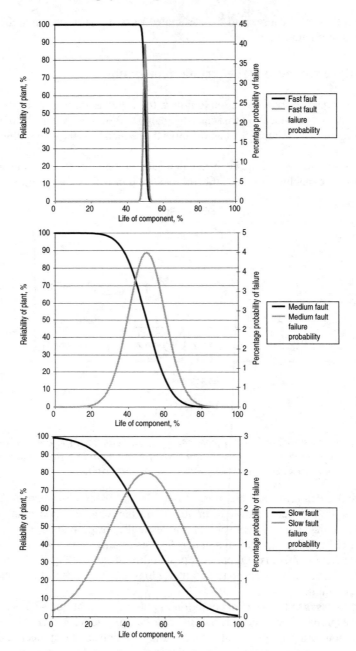

Figure 3.2   The failure sequence showing operability falling with time as a fault progresses. (a) Fast speed fault. (b) Medium speed fault. (c) Slow speed fault.

of impending failure and the ability to continue operating the machine before failure and maintain it to avoid failure. Therefore condition monitoring must concentrate on those root causes and failure modes that exhibit a failure sequence of substantial duration and our ability to detect its initiation and progress is crucial to successful condition monitoring.

3.4 Typical root causes and failure modes

3.4.1 General

It is important to distinguish between root causes, which initiate the failure sequence and can be detected by condition monitoring, and failure modes, which terminate it. After a failure, operators are used to tracing the sequence back from the failure mode to the root cause, in order to determine the true cause of failure. This is the process of root cause analysis (RCA) or asking why a failure occurred. On the other hand the designer of a condition-monitoring system must keep in mind the need to predict the reverse of that process, tracing how a failure develops, as shown in Figure 3.1. On this basis the authors propose the following as the most common root causes and failure modes in rotating electrical machines, based on the descriptions in Chapter 2 and the Appendix. They are similar to the root causes and failure modes identified by the IEEE survey [2] and by Thorsen and Dalva [3] and could be related to the more detailed analysis of failure modes developed by Bonnett and Soukup [4] for induction motors. It is surprising how few root causes and failure modes there are and the reader should note that these failure modes and root causes are generic and could be applied to many different subassemblies and components of the machine.

3.4.2 Root causes

These are outlined as follows

- defective design or manufacture,
- defective material or component,
- defective installation,
- defective maintenance or operation,
- ambient conditions,
- overspeed,
- overload,
- low cycle fatigue or shock load,
- high cycle fatigue or excessive vibration,
- component failure,
- excessive temperature
  - winding overtemperature,
  - bearing overtemperature.
- excessive dielectric stress, steady or transient,
- debris or dirt,
- corrosion.

3.4.3   Failure modes

These are outlined as follows

- electrical:
  - core insulation failure;
  - stator winding insulation failure;
  - rotor winding insulation failure;
  - brushgear failure;
  - slip ring failure;
  - commutator failure;
  - electrical trip.
- mechanical:
  - bearing failure;
  - rotor mechanical integrity failure;
  - stator mechanical integrity failure.

3.5   Reliability analysis

Component failure modes and the failure sequence described in Section 3.4 are controlled by statistical processes in the components, such as insulation degradation or degradation of metal components due to fatigue. These processes govern the transition from operation to failure exemplified by Figure 3.2. This process is the subject of detailed texts on reliability such as Billinton and Allen [5] and Caplen [6]. Each component failure mode will have its own probability density function that derives from its own physical processes. For example, the shape of the probability density function for winding insulation deterioration depends upon ambient and operating temperatures, deteriorating slowly as the insulation degrades with time. Whereas the probability density function for shaft fatigue would show a rapidly rising trend as fatigue cycles are accumulated.

The mathematical function $f(t)$ for a component failure mode probability density function should have the flexibility to represent a wide range of major failure modes, such as are present in most machinery. A flexible function that is available for this work is the Weibull function:

$$f(t) = \frac{\beta}{\theta} \left(\frac{t}{\theta} \right)^{\beta-1} e^{-(t/\theta)^\beta} \qquad (3.1)$$

The instantaneous failure rate, $\lambda(t)$, or hazard function of a component in a population of components can be obtained from $f(t)$ because

$$\lambda(t) = \frac{f(t)}{R(t)} \qquad (3.2)$$

where $R(t)$ is the reliability or survivor function of a population of components, which for a Weibull distribution would be given by

$$R(t) = e^{-(t/\theta)^\beta} \qquad (3.3)$$

Therefore

$$\lambda(t) = \frac{\beta}{\theta} \left(\frac{t}{\theta} \right)^{\beta-1} \tag{3.4}$$

which can be simplified to

$$\lambda(t) = \alpha \beta t^{\beta-1} \tag{3.5}$$

where β s a shape parameter and α is a scale parameter equal to $1/\theta^\beta$ and reduces to $1/\theta$ when $\beta = 1$ where θ is the MTTF of the component and $\approx$ MTBF. This expression is a very powerful mathematical tool to understanding the behaviour of components, subassemblies and complete machines from a reliability point of view.

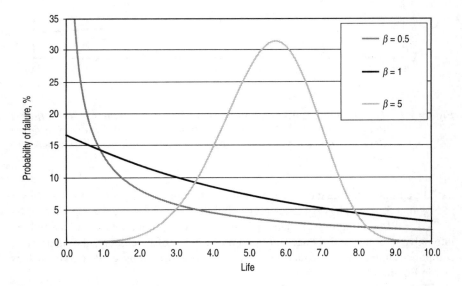

Figure 3.3   Variation in probability density function of failures to a component showing the effect of different degradation processes at work

Figure 3.3 shows the effect on the probability of failure of a component for different values of β, using (3.1) in this case life is expressed in arbitrary units. So for example the curve with $\beta = 0.5$ could describe the behaviour of a population of rolling element bearings whose probability of failure decreases progressively with time as the bearing and grease runs in.

On the other hand the curve with $\beta = 2$ or 5 could describe the behaviour of an insulation component subjected to increased degradation with time and temperature. Figure 3.4 shows the hazard function or failure rate using (3.5) for such components against life. The bearing component shows a failure rate decreasing steadily with time, whereas the insulation component shows an accelerating failure rate.

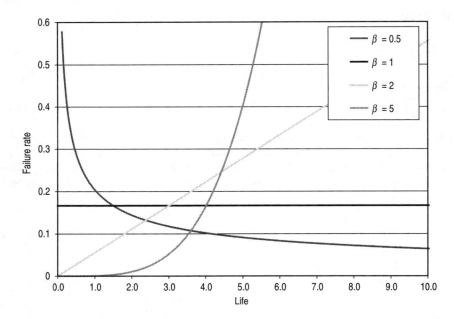

Figure 3.4 Variation in hazard function of a population of components for different underlying probability density functions showing the effect of different degradation processes on the deterioration of failure rate and therefore on the reliability

The complete failure rate of a population of components varies over time as shown in Figure 3.5. Each phase of the segregated failure rate curve can be represented by (3.5), for different values of shape parameter β. When components are assembled into subassemblies and thence into a complex piece of machinery, the aggregated component failure mode sequences results in a composite failure rate curve, such as that in Figure 3.5, derived from (3.5).

Figure 3.5 is a combination of three curves that represent three stages of operation of the population under consideration:

- *Early life* ($\beta < 1$): this stage is also known as infant mortality or burn-in, where the failure rate falls as the components prone to early failure are eliminated.
- *Useful life* ($\beta = 1$): this stage represents normal operation of the system, when the failure rate is sensibly constant over an extended period of time.
- *Wear-out* ($\beta > 1$): this stage represents increasing failure rate of the system, occurring towards the end of its useful life, due to deterioration of the individual components that start to fail in increasing numbers.

However, pieces of machinery do not always follow the typical bathtub form. The authors have noted how complex electronic equipment does not often exhibit infant mortality symptoms. This is because such subassemblies are subjected to

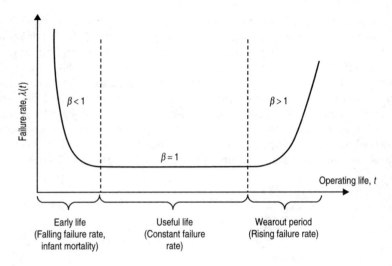

Figure 3.5   *Aggregate hazard function or failure rate for a population of components or subassemblies forming a complete piece of machinery, known as the bathtub curve showing the effect of the shape function β at the beginning and end of machinery life*

automatic and accelerated life testing on completion of manufacture, before delivery, specifically designed to minimise the number of early-life failures.

In another way, electrical insulation systems do not tend to show a low, constant, mid-life failure rate but a rather gradual worsening in failure rate over the whole life, as the insulation material steadily deteriorates with ageing, as shown in Figure 3.4 with $\beta = 2$. However, the hazard function of the bathtub curve is instructive for those engaged in condition monitoring because they demonstrate the character of failure modes, which we want the condition monitoring to detect, and the different phases of machinery life.

The low, constant, mid-life failure rate at the bottom of the bathtub is achieved partly by good design and manufacture of machinery components but its base value can be reduced and duration extended by maintenance and in turn by monitoring. So the shape of the bathtub for a complex engineering plant is dependent upon the maintenance and monitoring regime that is adopted as set out in Chapter 1.

3.6   Machinery structure

The structure of a machine is an important factor in the cause-and-effect diagram and in the aggregation of failure mode probability density functions, as described in the previous section. We may consider that structure to be simply the assembly structure of individual components into subassemblies and then the aggregation of those subassemblies into a complete machine. This could be exemplified by the exploded diagram of the machine and was discussed in Chapter 2 (see Figure 2.10 for example).

Billinton and Allen [5] make clear that the relevant structure is not assembly but that due to the reliability dependence of components, where the failure of one component in the structure may not incur the failure of the machine, for example due to redundancy. This affects the makeup of the cause-and-effect diagram, which is the structure relevant to condition monitoring, and is governed by causality rather than assembly.

The structure for a subassembly can be built up from the cause-and-effect diagrams of individual components to give a cause-and-effect diagram for the subassembly as shown in Figure 3.6. The assembly structure for a typical electrical machine is shown in Figure 3.7 and may be adequate in some cases but is not necessarily correct for the assessment of failure modes. Compare Figure 3.7 to the diagram in Figure 2.10.

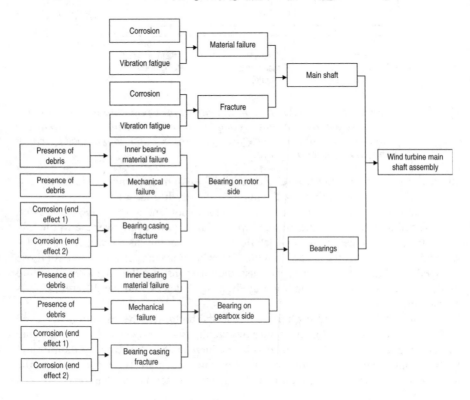

Figure 3.6   Example of a cause-and-effect diagram for a simple subassembly, the main shaft of a wind turbine

A simple example of the two aspects of this issue could be the lubrication pumps of a large turbo generator. It is customary to introduce redundancy into this plant by providing two off, 100 per cent-rated, main and standby, AC motor-driven lubrication pumps with a third 100 per cent-rated, DC motor-driven pump to provide emergency power from a battery when AC power fails. In this case it is not legitimate simply to add the cause-and-effect diagrams in series or to add the failure mode probability

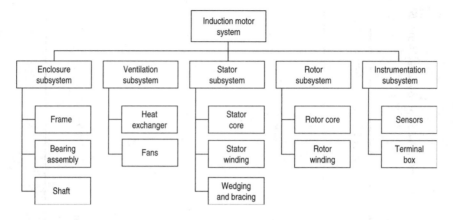

Figure 3.7 Example of a typical electrical machine structure diagram (see Figure 2.10)

functions of the three pumps, since a large turbo generator can operate with any one out of three pumps functioning. The correct approach is to consider the three systems in parallel and take account of the probability density functions in the same way.

3.7 Typical failure rates and MTBFs

The approach of this chapter can be put into context by considering the MTBFs of typical electrical machines. Such data can be notoriously difficult to find but some information, particularly about MTBF, is available from reliability surveys, mostly conducted in the US under the auspices of the IEEE [2]. Information about the life of electrical machines has also come to light from experience in the defence industry, where reliability predictions are a contractual requirement of equipment purchase [7]. Finally information is available from the wind industry about generators being fitted to the wind turbines being installed in increasing numbers, where fault data is recorded sufficiently frequently to deduce the life curves of typical electrical machines and their reliability [8].

The MTBF can be a deceptive quantity. It is intended to represent the prospective life of the machine, assuming it has a constant failure rate, as shown in the constant failure rate region of Figure 3.5. Then one could consider that there would be a 50 per cent probability of failure before the MTBF and 50 per cent probability of failure afterwards. However, machines can have a failure rate that improves with time and then it is possible that a higher proportion of failures could occur before the MTBF. However, a well-maintained machine can expect to be operating in the constant failure rate region, in which case MTBF gives a good indication of prospective life, which engineers can appreciate without being overwhelmed by any statistical interpretation.

Table 3.1 extracts data from a number of surveys of electrical machines and gives the failure rates and MTBFs for a range of machines, showing a remarkable degree of

Table 3.1 A list of typical measured failure rates and mean time between failures (MTBFs) for electrical machines obtained from the literature

Machine type	No. of machines in survey	Machine years in survey	No. of failures in survey	Failure rate, failures/machine/year	MTBF, hours	Source of data
Large steam turbine generators	Not known	762	24	0.0315	273 750	[2,9]
Induction motors 601–15 000 V	Not known	4 229	171	0.0404	216 831	[2,9]
Motors greater than 200 hp, generally MV and HV induction motors	1 141	5 085	360	0.0707	123 735	[2,10]
Motors greater than 100 hp, generally MV and HV induction motors	6 312	41 614	1 474	0.0354	247 312	[11]
Motors greater than 11 kW generally MV and HV induction motors	2 596	25 622	1 637	0.0639	137 109	[12]
Wind turbine generators <2 MW	643	5 173	710	0.0400	219 000	[8]

consistency with MTBFs ranging from 18 to 33 years, if a year is assumed to contain 8 766 hours operation. The table also gives an idea of the significance of each survey by noting the number of machines surveyed and the number of failures recorded. It should be noted that the large surveys are dominated by induction motors because of their ubiquity.

The distribution of failures within the structure of the machine is also important because it should guide the direction in which condition monitoring is applied. Table 3.2 gives an analysis of failures based on the literature and three important areas of the machine are identified: stator related, primarily the winding; rotor related, including slip rings and commutators; and bearings. The remainder of failures are grouped as other. The data comes from different surveys and while these surveys are not necessarily complementary they are substantial and do show a consistent aspect of failure areas in descending order of importance

- bearings,
- stator related,
- rotor related.

The relative importance of failures is affected by the size, voltage and type of machine under consideration. In particular the relative weighting of stator to rotor winding failures does depend upon the type and size of machine under consideration. For example small, low-voltage induction machine failures, exemplified by the first two columns of Table 3.2, are dominated by bearings, as low-voltage windings experience very few faults. Smaller machine bearings are usually rolling element and their reliability depends heavily on the standard of maintenance. Induction machines show a much lower number of rotor winding or squirrel cage faults compared to stator winding faults, because of the ruggedness of cage construction. But larger, high-voltage machines, exemplified by the next three columns of Table 3.2, receive a higher proportion of failure modes on the stator winding, due to dielectric stress and vibration root causes, and this can rise to be of a similar significance to bearing faults. Large machine bearings are also usually of sleeve construction and with constant lubrication are generally more reliable than bearings in smaller machines. Table 3.2 also shows that other component failure modes, for example in the cooling system, connections and terminal boxes, become more significant on larger machines.

An interesting aspect of this analysis is the attention that is being paid to different machine failure mechanisms in the published literature. The authors have used the search engine IEEEXplore to consider IEEE and IEE journal papers only in the period 1980 to date and searched the metadata under the following headings:

- broken bars in induction motor rotor cages: 35 papers;
- discharge activity in medium- and high-voltage stator windings in motors and generators: 9 papers;
- stator winding faults excluding discharge activity in motors and generators: 19 papers;
- bearing; faults in electrical machines: 17 papers.

Table 3.2 The distribution of failed subassemblies in electrical machines obtained from the literature

Subassemblies	Predicted by an OEM through failure modes and effects analysis techniques, 1995–1997*	MOD survey; [7]	IEEE large motor survey, 1985; [10]	Motors in utility applications; [11]	Motor survey offshore and petrochem; [12]	Proportion of 80 journal papers published in IEEE and IEE on these subject areas over the past 26 years
Types of machines	Small to medium low-voltage motors and generators <150 kW, generally squirrel cage induction motors	Small low-voltage motors and generators <750 kW, generally squirrel cage induction motors	Motors greater than 200 hp generally medium- and high-voltage induction motors	Motors greater than 100 hp generally medium- and high-voltage induction motors	Motors greater than 11 kW generally medium- and high-voltage induction motors	All machines
Bearings	75%	95%	41%	41%	42%	21%
Stator-related	9%	2%	37%	36%	13%	35%
Rotor-related	6%	1%	10%	9%	8%	44%
Other	10%	2%	12%	14%	38%	—

* Private communication from Laurence, Scott & Electromotors Ltd

The spread of this collection of 80 publications is shown distributed alongside the relevant failure areas in Table 3.2. It shows that more publishing effort has gone into the study of the less prevalent rotor faults than into the more prevalent bearing faults.

Induction motor rotor cage faults can be detected through perturbations of the airgap magnetic field (see Chapter 9). The distribution of journal papers towards those faults is probably due to the fact that the airgap field is scientifically interesting, analytically tractable and cage induction motors are very numerous.

Bearing faults, on the other hand, involve a more complex interplay of mechanical and electrical physical effects. However, bearing faults lead to airgap eccentricity and the effect on the resultant magnetic field is also tractable, although more complex than the effect of rotor cage faults. Some of the lessons learnt from rotor cage fault detection, combined with the study of the effect of eccentricity on the airgap magnetic field, can be applied to the detection of bearing faults and this topic is returned to in Chapters 8 and 9. It is interesting that the study of the effects of eccentricity in induction motors, the most numerous of electrical machines, can also be elicited from the literature survey given earlier and that 36 papers have addressed the subject since 1980. Some of these papers address issues of noise and speed control but the number demonstrates the attention that is beginning to be focused on this important issue.

3.8 Conclusion

This chapter has shown that causality must be the guiding principle when applying condition monitoring to electrical machines. First, the operator must be aware of the failure modes and root causes for the machine being considered for condition monitoring. Second, causality must be traced through prospective failure sequences, possible by the use of cause-and-effect diagrams. The probability of failure of a component of a machine can be described by a probability density function for that component. The resultant hazard function curves for each component in a subassembly can be aggregated to give a prospective life curve for that subassembly and can then be aggregated to give the life curve of the machine. The aggregation of cause-and-effect diagrams and hazard functions needs to be done with care, taking account of the assembly structure of the machine and the reliability dependence of components. From the aggregate hazard function a model for the complete machine could be derived. Third, condition monitoring needs to address the root causes that have a slow failure sequence to the failure mode and this information can be derived from knowledge of the failure modes in operation in an electrical machine.

In Chapter 1 we proposed three different courses of maintenance action:

1. breakdown maintenance, implying uncertainty and a high spares holding;
2. fixed-time interval or planned maintenance, implying extended shut-down or outage periods and lower availability;
3. condition-based maintenance, implying low spares holding and high availability.

Real engineering plant demands a flexible combination of maintenance regimes based on the above. This chapter has shown that by addressing failure modes that are slow to mature, condition monitoring could significantly affect the detection of faults before they occur. Condition monitoring can then form part of planning fixed-time interval maintenance (2) and will be at the heart of maintenance for method (3).

Prior to determining the type of equipment and frequency of monitoring, some consideration should also be given to whether the cost of implementing such a program is justified. Factors involved in this decision may include

- replacement or repair cost of the equipment to be monitored,
- criticality of the plant to safe and reliable operation,
- long-term future of the facility in which the equipment is installed.

Assuming that consideration of these points indicates that the expenditure is justified, the following points will aid in determining how the condition-monitoring programme should be implemented

- design,
- manufacture,
- installation,
- operation,
- maintenance.

In the majority of cases, the end users of rotating machines are dealing with existing plant and will have to consider all five of these points. This task is complicated because often there is very little design information, the manufacturer may no longer exist and installation, operation and maintenance records may be incomplete. The paucity of information can lead to the assumption that the age of the equipment indicates the need for more rigorous monitoring. However, this is not always the case since the refinement of design tools and the constant pressure on manufacturers to reduce costs has, in some cases, led to decreased design margins.

The practicability of these approaches for electrical machines depends upon the application of the machine and the engineering plant it serves. That is the basis of this book. In the next two chapters we describe the instrumentation and signal processing needs for condition monitoring.

3.9   References

1. US Department of Defense. MIL-STD-712C Military Standard Procedures for Performing a Failure Mode, Effects and Criticality Analysis. Washington DC: US Government; 1980.
2. IEEE Gold Book. Recommended Practice For Design of Reliable Industrial and Commercial Power Systems. Piscataway: IEEE Press; 1990.
3. Thorsen O.V. and Dalva M. Failure identification and analysis for high-voltage induction motors in the petrochemical industry. *IEEE Transactions on Industry Applications* 1999; **IA-35**: 810–18.
4. Bonnett A.H. and Soukup G.C. Cause and analysis of stator and rotor failures in three-phase squirrel-cage induction motors. *IEEE Transactions on Industry Applications* 1992; **IA-28**: 921–37.

5.   Billinton R. and Allan R.N. *Reliability Evaluation of Engineering Systems – Concepts and Systems*. 2nd edn. New York: Plenum; 1992.
6.   Caplen R.H. *A Practical Approach to Reliability*. London: Business Books; 1972.
7.   Tavner P.J. and Hasson J.P. Predicting the design life of high integrity rotating electrical machines. Presented at the 9th IEE International Conference on Electrical Machines and Drives (EMD), Canterbury, 1999.
8.   Tavner P.J., van Bussel G.J.W. and Spinato, F. Machine and converter reliabilities in wind turbines. Presented at the 3rd IET International Conference on Power Electronics Machines & Drives (PEMD), Dublin, 2006.
9.   Dickinson W.H. IEEE Reliability Working Group. Report on reliability of electric plant, part I, II and III. *IEEE Transactions on Industry Applications* 1974; **IA-10**: 201–52.
10.   O'Donnell P. IEEE Reliability Working Group. Report of large motor reliability survey of industrial and commercial installations, part I, II & III. *IEEE Transactions on Industry Applications* 1985; **IA-21**: 853–72.
11.   Albrecht P.F., Appiarius J.C. and Sharma D.K. Assessment of the reliability of motors in utility applications – updated. *IEEE Transactions on Energy Conversion* 1986; **EC-1**: 39–46.
12.   Thorsen, O.V. and Dalva M. A survey of faults on induction motors in offshore oil industry, petrochemical industry, gas terminals, and oil refineries. *IEEE Transactions on Industry Applications* 1995; **IA-31**: 1186–96.

Chapter 4
Instrumentation requirements

4.1 Introduction

The development of a condition-monitoring system involves the measurement of operating variables of the plant being monitored, and interpretation as well as management of the acquired data. In many plants, measurements are already used for control and protection objectives. From a cost point of view, it would be ideal if such measurements could also be used for condition monitoring. However, knowledge of the limitations of the existing measurements and understanding of the difference between the objectives of control, protection and condition monitoring is important, as will be seen later in the book.

One feature of condition monitoring, described in Chapter 3, is to detect impeding faults at an early stage, capturing weak signatures in measurements that are usually mixed with noise. Furthermore, some impeding faults may manifest themselves in non-electrical variables that are not normally used in control or protection. Before getting into the details of the instrumentation elements, we need to view a condition-monitoring system from a higher level where the functionality of different parts of the system can be more clearly described. By doing this, it should be possible to identify the common elements of a condition-monitoring scheme, irrespective of the system detail. In essence we are saying that an engineer examining occasional meter readings, with a view to producing an operational and maintenance strategy, is involved in a procedure that is analogous to a sophisticated condition-monitoring system.

In view of this analogy, we believe that the following tasks are essential elements in a condition-monitoring system (Figure 4.1).

1. The measurement or transduction task (sensing of primary variables).
2. The data acquisition task (conversion of sensed variables into digital data in condition-monitoring system).
3. The data processing task (identifying of information buried in data).
4. The diagnostic task (acting on processed data).

Tasks (1) and (2) are usually carried out while the plant being monitored is operating. Tasks (3) and (4) can be performed off-line and the results of these tasks may not be fed back immediately to the plant operators. This is a non-intrusive approach of condition monitoring. In contrast, it is possible to use an intrusive approach in which a signal is injected into the plant being monitored and the response is used to control the plant status. This is particularly attractive in plant involving power electronic

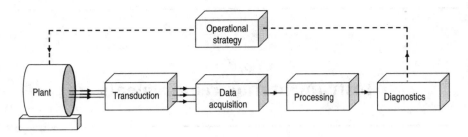

Figure 4.1   The monitoring tasks

converters that generate monitoring signals and are very fast acting. For the time being, we will confine our description to the non-intrusive approach.

In each of these four tasks human intervention is possible and often desirable. Indeed in some cases some of these tasks may be entirely carried out by the operator. Let us return to the analogy of the engineer collecting their meter readings, and try to identify the tasks. The meter deflections or the numbers displayed on the panels will naturally be in response to measurements made elsewhere in the plant under monitoring, and may be picked up by current transformers, voltage transformers, thermocouples or accelerometers, for example. This is the transduction task, and is normally performed automatically as it is difficult for the human operator to act in anything but a qualitative way in fulfilment of this task. That is, they may be able to tell if one piece of plant is hotter than another but little else. Having that said, it often relies on a human operator to ensure that the result of the transduction task is reliable.

The data acquisition task could be simplified to the action of the engineer in writing down the series of readings from each meter, together with information regarding the time, location, loading and ambient conditions perhaps, of the plant under monitoring. In our example this is wholly under human control. The frequency of such readings, and how long the observations are made, will depend on the characteristic of the plant.

The processing task corresponds to the analysis of the readings in the same way. It may be considered appropriate to average several of the readings for example, and to present them in a way that allows easy comparison with other data. This phase can vary greatly in complexity, and it is this task that demands the application of significant computing power in automatic systems. We shall return to many of the more common processing techniques later merely noting at the moment the use of methods such as spectral analysis, time averaging, and auto- and cross-correlation.

The diagnostic task operates on the results of the processing task in order to recommend actions that will hopefully result in improved operational readiness and performance, and improved maintenance scheduling. Again, returning to our simple analogy, on the basis of examining the average meter readings and comparing them with the manufacturer's operational limits, say, the engineer decides for the next plant shutdown which items of machinery must be overhauled or replaced. The engineer arrives at this judgement on the basis of their experience and the data available to them as a result of collecting and processing the meter readings, and having access to the manufacturer's data on operational limits. For the manufacturer to set the operational limits, a relationship between the plant condition and the signature during operation should have already been established. Understanding of the mechanism of fault development is particularly useful in determining such threshold values. The diagnostic task is still most often carried out with significant input from the human operator, but with the development of knowledge-based expert systems or other forms of artificial intelligence. Full automation of this phase is being realised.

Recent development of communication technology; for example, in fibre-optics and semiconductor sensing devices, allows the data acquired to be sent to a central point or to the manufacturer where they are processed and analysed. With the current trend of restructuring in many industrial sectors, the responsibility of plant condition monitoring has also started to change; operators, asset owners and manufacturers may all have parts to play with the manufacturers taking more responsibility than they traditionally have. Depending on its application, a plant can be equipped with condition-monitoring functionalities as a selling point. As manufacturers cannot schedule plant shutdowns, the central position of the operator could hardly be replaced. Here we are mainly interested in the technical aspects of condition monitoring. The trend towards thorough life costing of plant means that manufacturers are more closely involved with the operator in managing plant beyond the warranty period and this opens up an increased market for service work, which can be sustained by condition-monitoring activity. In Figure 4.1, we summarise, in schematic form, the four tasks associated with condition-monitoring activity, and indicate the obvious step of closing the loop as an aid to improve the operational performance.

We must now examine the first two of the four condition-monitoring tasks in some detail, both in a general way and as they apply to electrical machines. The data processing task will be examined in Chapter 5 while the diagnostic task will be delayed to later in the book. At all times we must be mindful that the monitoring activity is aimed at reaping the benefits to be gained by closing the loop identified in Figure 4.1.

4.2 Temperature measurement

Temperature is widely monitored in electrical drives and generators. For example, it is common to find temperature sensing used to monitor specific areas of the stator core and the cooling fluids of large electrical machines such as turbine generators. Such measurements can only give indications of gross changes taking place within the machine but they are extremely effective if mounted and monitored in carefully selected sites. When temperature measurement is combined with information about the loading and ambient conditions of the machine, it provides valuable monitoring information. Bearing temperatures are commonly monitored on drive trains and, together with vibration sensing, temperature measurement provides the standard approach to the assessment of the condition of these elements. The typical techniques and sites for temperature measurement in an electrical machine are described in Chapter 8. Here we describe the three principal methods of measuring temperature electronically

- resistance temperature detection,
- thermistors,
- thermocouples.

Each type has acceptable areas of application, which we will now briefly investigate.

The resistance temperature detector (RTD) is also called the resistance thermometer, which uses the resistance change of a metal to indicate temperature change. Platinum and nickel are usually used because they are retardant to corrosion and because they have sensitive resistance versus temperature characteristics, as shown in Figure 4.2 where R_0 is the resistance of the device made of the metal at 0 °C [1]. An RTD can be made by either winding insulated metal wire around a thin cylindrical former or by evaporating a thin coating of the metal on an insulating, usually ceramic, substrate. Figure 4.3 shows a commercial sensor enclosed in a thin stainless

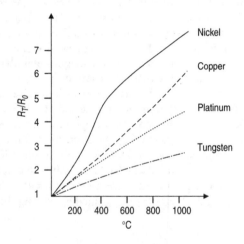

Figure 4.2   Typical temperature-resistance characteristics of metals

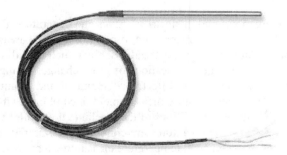

Figure 4.3   A resistance temperature detector device [Source: Labfacility Ltd.]

steel tube for protection. Such devices are widely used for gas and liquid temperature measurement applications up to 1 000 °C and are the usual choice for electrical
machine manufacturers for insertion between winding conductors in machine slots.

RTD devices are constructed generally to a base resistance of 100 Ω at 0 °C and
have the advantage of being linear over a wide operating range. They have very good
accuracy and precision but have a relatively low sensitivity. They are generally used in
a four-wire configuration of Wheatstone bridge, but two-wire or three-wire operation
is also possible. The circuit converts the change of resistance into the change of output
voltage.

Figure 4.4 shows a Wheatstone bridge circuit for an RTD device made of
platinum. The RTD device has the following temperature–resistance relationship:
$R_T = R_0(1 + \alpha T)$ where R_T is the resistance in ohms (Ω) and T is the temperature
in degrees celsius; $\alpha_r = 0.00385/°C$, $R_0 = 100Ω$ and the power supply to the circuits is $V_s = +15\ V$. Ignoring the zero adjustment resistance, Table 4.1 shows the
output voltage calculated for two resistance values: $R = 5\,530Ω$ and $R = 48Ω$. The
overall sensitivity and linearity between 0–100 °C are then derived. The sensitivity
is defined as the ratio of change of output to the change of input while the linearity
is defined as the deviation from the straight line between the two end points of the
whole range, that is from 0–100 °C. It is clear that increasing the value of R causes
the sensitivity to be reduced but the linearity to be improved. The self-heating error
is reduced as the value of R is increased. The Wheatstone bridge arrangement allows
the lead resistance to be easily compensated as shown in the figure.

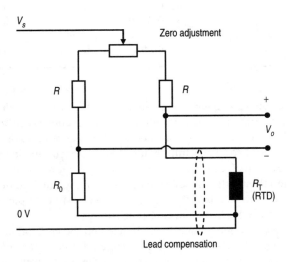

Figure 4.4   Wheatstone bridge circuit for a resistance temperature detector

Temperature measurement using thermocouples is based on the well-known See-
beck effect whereby a current circulates around a circuit formed using two dissimilar
metals, when the metal junctions are held at different temperatures. If a break is made

Table 4.1   Response of a Wheatstone bridge resistance temperature detector circuit

$T(°C)$	$R_T(\Omega)$	$V_0(V@R = 5\,530\Omega)$	$V_0(V@R = 48\Omega)$
0	100	0	0
20	107.79	0.0204	0.24
40	115.54	0.0406	0.46
60	123.24	0.0606	0.66
80	130.89	0.0804	0.84
100	138.5	0.1	1.0
Sensitivity		0.001 V/°C	0.01 V/°C
Linearity		0.6°C error	6°C error

Table 4.2   Ranges, outputs and sensitivity of thermocouple junction materials

Type	Junction materials	Range (°C)	Output at 100 °C (mV)	Sensitivity at 30 °C (mV/°C)
E	chromel/constantan	−279/1000	6.317	0.0609
J	iron/constantan	−210/1200	5.268	0.0517
K	chromel/alumel	−270/1370	4.095	0.0405
N	nicrosil/nisil	−270/1300	2.774	0.0268
T	copper/constantan	−240/400	4.277	0.0407

in one of the wires then a voltage is generated across the break, which increases with the temperature difference between the junctions. One junction is held at a temperature that can be easily measured to allow the so-called reference junction compensation. By doing this the need for a carefully controlled reference junction temperature is avoided and a device with effectively only a single junction results [2]. Some typical junction materials and their associated operating ranges and outputs are given in Table 4.2, derived from standard thermocouple tables. Multiple thermocouples can be combined to form a thermopile of higher sensitivity.

Figure 4.5 shows thermocouples being bonded on the coils of an air-cored axial flux permanent magnet motor in an electric vehicle for condition monitoring. It is essential to use the correct leads when extending the wires to the point where the output voltage is measured. As the thermocouple output is already in the form of a voltage, there is no need to use additional circuit such as a Wheatstone bridge for signal conversion. However downstream devices are still generally required for amplification and reference junction compensation. Figure 4.6 shows a reference junction temperature compensation circuit using an LM35 integrated circuit whose

Figure 4.5   Thermocouples to monitor coil temperature in a permanent magnet drive motor

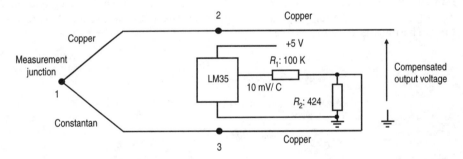

Figure 4.6   Reference junction compensation for a thermocouple

output voltage is 0 V at 0 °C and increases at a rate of 10 mV/°C. A voltage divider is included to cause an approximately 40 μV/°C voltage gradient across resistor R_2 that forms part of the measured output voltage. The voltage across R_2 will then compensate the voltage developed across the cold junction marked as '3' in Figure 4.6, assuming that the cold junction is at room temperature, around 18–23 °C. When appropriate, the sparking welding technique can be used to cause a thermocouple junction directly on a metallic surface.

The working life of thermocouples depends on the working temperature and the physical dimensions of the junction materials, but copper/constantan and chromel/alumel, which are most widely used in electrical machine condition monitoring, can be expected to survive for many years. Depending on the quality of the signal conversion circuit, the performance of RTD temperature transduction can be very good. It is, however, relatively expensive. Thermocouples usually offer slightly reduced performance at much more modest cost. We noticed that the changes of output involved in both types of temperature transducer were small.

Another type of temperature transducer we wish to briefly mention is the thermistor, which provides a coarse but very sensitive response. Thermistors are manufactured from blends of metal oxides of cobalt, iron, titanium and nickel, which are fired like clays into small discs or beads that may be encapsulated in resin or enclosed in protective brass tubes. They are also available in the form of washers for easy mounting. Figure 4.7 shows a range of commercial thermistors. Thermistors exhibit a large change in resistance as a function of temperature. In addition to sensitivity, they also have the advantages of high stability, fast response and very small physical size. Typically a glass-bead device will have a diameter in the order of 0.25 mm, while flake thermistors are produced in thickness down to 0.025 mm with cross-sections of 0.5×0.5 mm. The size advantage means that the time constant of thermistors operated in sheaths is small. But the size reduction also decreases the heat dissipation capability and so makes the self-heating effect of greater importance. Thermistors are generally limited to 300 °C, above which level the stability reduces. They do not provide a linear output and this must be taken into account. For this reason, it is also difficult to control the consistency between samples of thermistors, implying that the precision achievable is relatively low. A thermistor is usually used as a logic-switching device to operate an alarm when the temperature exceeds a preset limit. Thermistors are inexpensive.

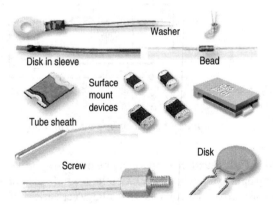

Figure 4.7   Commercial thermistors [Source: EPCOS Inc. and Tyco Electronics Corp.]

Figure 4.8 shows four typical temperature-resistance characteristics of thermistors. The nominal values of resistance 10 kΩ, 20 kΩ, 30 kΩ or 100 kΩ are defined for a room temperature of 25 °C. We note that the vertical axis is in logarithmic scale implying that the change of resistance with temperature is indeed

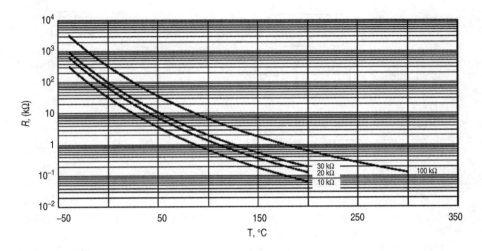

Figure 4.8   Thermistor characteristics

significant. Since the accuracy of the device is hardly high and the characteristic may also slightly drift after long-time thermal cycling, the signal conversion scheme adopted for a thermistor is usually quite simple; in most applications, a voltage divider will be more sensible than a Wheatstone bridge circuit, helping to keep the overall cost down.

In recent years, semiconductor temperature transducers in the form of integrated circuits have found more applications [1]. Their design results from the fact that semiconductor diodes (*p–n* junctions) have temperature sensitive voltage-current characteristics. With external power supply and internal feedback mechanisms, transistors in the integrated circuit can be made to produce an output current or voltage that is proportional to the absolute temperature. A voltage is developed as the current flows through a fixed resistor. The use of integrated circuit temperature sensors is limited to applications where the temperature is within a −55 °C to 150 °C range. Their errors are typically ±3 per cent, quite low indeed. They are very inexpensive and therefore have been used in large quantities to monitor the temperature distribution of equipment with large dimensions such as pipes and cables. We have described reference junction compensation for a thermocouple. Such integrated circuits can also be used for that purpose.

Other techniques such as quartz thermometers, fibre-optic temperature sensing and infrared thermography can be used to achieve high accuracy, fast response or provide non-contact and spark-free solutions. They are generally more expensive than the techniques described earlier and have not been widely used in condition monitoring of electrical machines. However, in cases where the value of the monitored equipment greatly exceeds the cost of the transducers, such techniques are now increasingly being used. For instance, fibre-optic temperature sensing has been applied to identify hot spots in switchgear, transformers and more recently synchronous generators [3,4].

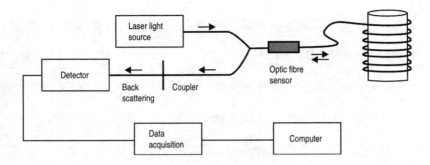

Figure 4.9   Principle of fibre-optic temperature sensing

For this reason, we now outline the principle and characteristics of such a technique, indicating its advantages and limitations.

Figure 4.9 shows the basic concept of fibre-optic temperature sensing. When a laser pulse is injected into the optic fibre, it is subject to scattering as it travels and the back scattered pulse is returned as shown in the principle of optical time domain reflectometry [4,5]. The back-scattered light consists of a Rayleigh component, a Brillouin component and a Raman component, which differ in wavelength. Thermally influenced molecular vibration causes the Raman-scattered component to change and therefore it is sensitive to temperature. According to the time when the signatures in the back-scattered light are received, the temperature along the optical fibre can be detected. Therefore fibre-optic temperature sensing can be a naturally distributed method. Spatial resolution is typically in the 1 m range in terms of the fibre length. By winding the optic fibre around the body whose surface temperature is to be measured, better spatial resolution in terms of physical body dimensions can be achieved [6]. Using a technique called fibre Bragg grating, distributed temperature sensing can be multiplexed with strain and displacement measurements in the same fibre-optic system, as to be described later in this chapter. The temperature resolution is typically less than 0.1 °C in the range of zero to a few hundred degrees celsius. The response time of fibre-optic temperature sensing is around 2 s.

4.3   Vibration measurement

4.3.1   General

In Chapter 8 we will discuss at length the techniques available and the applicability of vibration monitoring. This is particularly useful for monitoring the condition of components in the mechanical drive train of electrical machines, such as gearboxes, shaft couplings, bearings and rotor unbalance. More recent work has extended that to the extraction of condition information from electrical variables such as current, vibration sensing (including acoustic noise) remains a most important monitoring technique for the operators of electromechanical plant [7]. Its use is extremely widespread and has reached a high degree of sophistication. It revolves around the measurement of three quantities that are related by numerical integration or differentiation

- displacement,
- velocity,
- acceleration.

Which quantity one should measure depends on the size of the plant being monitored, and the frequency range in which one is interested. Generally, machines of similar type and size have a more or less constant vibrational velocity. Also as the vibration frequency increases it is likely that displacement levels will fall but acceleration levels will rise. This suggests that with increasing frequency it is better to progress from a displacement device to a velocity transducer, and ultimately to an accelerometer. As a guide, the approximate frequency ranges of application are as shown in Figure 4.10. It is useful to appreciate that the process of numerical integration attenuates measurement noises while numerical differentiation amplifies them.

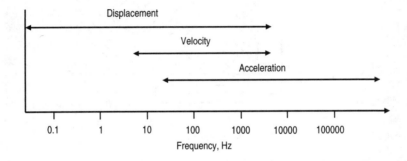

Figure 4.10   Normal frequency ranges of applicability for vibration measurements

Care must be exercised, however, when monitoring systems that have small moving masses, for in such circumstances transmitted forces may be small and the mass of the accelerometer may be significant compared to the body mass. Here, displacement will usually provide the best indication of condition. We can now characterise vibration transducers according to the quantity they measure.

4.3.2   Displacement transducers

Here we shall confine our interests to non-contacting displacement probes or proximeters. This excludes some capacitive and low-frequency inductive devices such as the LVDT (linear variable differential transformer). One type of non-contact device that is used in condition monitoring operates by using a high-frequency current/voltage source to generate an electromagnetic field at the probe tip. The system energy, due to eddy currents in the target, is dependent upon the local geometry of the area surrounding the probe tip. If the system energy changes (for example, when the target surface, which is ferromagnetic and electrically conductive, moves with respect to the probe) then the system energy also changes. This change is readily measured as the change of voltage/current in the high-frequency excitation circuit is related to the displacement of the target surface from the probe tip. It should be noted that such systems measure the relative motion between the probe and the target, hence the vibration of the housing in which the probe is mounted is not readily measured by this technique. Figure 4.11 shows the operation of a proximity probe.

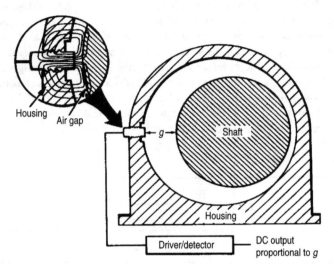

Figure 4.11   Operation of a proximeter probe

Sensitivities in the order of 10 mV/μm displacement are easily achievable with displacement probes, and they find wide application in situations where heavy housings ensure small external movements. The measurements of shaft eccentricity and differential movements due to expansion are therefore most easily achieved using displacement transducers. The same principle has been used to develop devices that effectively measure rotational displacement or speed by sensing the passage of keyways on the shafts. These are sometimes called key phasors.

As mentioned previously, displacement is most effectively measured at the lower frequencies even though the frequency range of eddy current systems can extend above 10 kHz. They are relatively robust transducers and the driving and detection circuits are straightforward. Essentially the high-frequency signal applied to the probe is modulated by the passage of the target, and the demodulated signal used as the measurement quantity. Consideration of Figure 4.11, which illustrates the basic displacement measurement principle, shows us that the output of the system will depend not only on the displacement between the probe and the target, but also on the material from which the target is made. This is because the eddy current reaction of the target, and hence the system energy, is dependent upon the conductivity and/or permeability of the material. Proximity probes must therefore be calibrated for each target material. Care must also be taken when mounting the probe to ensure that electrically conducting and magnetic surfaces around the probe tip do not cause unnecessary disturbance of the applied high-frequency field and that the target surface is smooth with no surface or magnetic disturbances.

Another non-contact device uses a fibre optic, which detects the light reflection and compares it with the injected light to determine the distance between the reflection surface and the optic-fibre end. This is illustrated in Figure 4.12. The result is converted into a voltage signal using a photodiode in the receiver. The contact-free measurement range is up to 0.3 mm with a typical resolution of 0.01 mm. There are other fibre-optic displacement sensors based on fibre Bragg grating that can have a larger measurement range. However they usually require contact to the target surface; temperature compensation may also be necessary as the displacement is translated into strain for measurement [8]. In general, fibre-optic devices are intrinsically immune to electromagnetic noise problems.

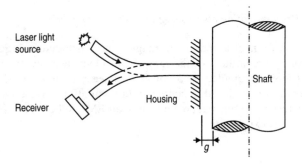

Figure 4.12   Principle of fibre-optic proximetry

4.3.3   Velocity transducers

Rotational speed in electrical machine systems is usually measured for the purpose of control. Optical encoders are now used as common practice. They effectively derive speed information by counting the position rotated per unit time. The optical encoder, which when in use is attached to the rotating shaft, carries a disk cut in slits. A light source and a receiving device are aligned on both sides of the disk. Light pulses are caused on the receiver, which are converted into electrical pulses by a photo-sensitive transistor. Such an encoder is very suitable to interface with a digital control system. The angular resolution is typically 0.18° corresponding to 2 000 pulses per revolution. Linear encoders are commercially available. However, application of such optical encoders for vibration measurement has been limited to the very-low-frequency range because position variation is hardly enough when the vibration frequency is high. High-precision electromagnetic resolvers can also be used to obtain rotational information from a drive shaft.

For vibration measurement, velocity is most usefully sensed in the frequency range from 10 Hz to 1 kHz. This is usually achieved in an analogue manner by designing a spring mass system with a natural frequency less than 10 Hz, and letting

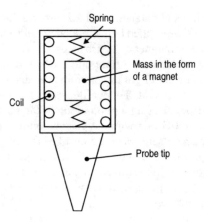

Figure 4.13   Principle of an electromagnetic velocity probe

the mass take the form of a permanent magnet, as illustrated in Figure 4.13. The device is responsive regardless of the range of displacement involved in vibration. The magnet is then surrounded by a coil that is securely attached to the housing. Whenever the housing is placed in contact with a vibrating surface the housing and coil move with respect to the magnet and cause an EMF to be induced in the coil, in accordance with the expression

$$e = \ell B v \tag{4.1}$$

where e is the induced EMF, ℓ the effective length of the conductor in the coil, B the radially directed flux density, which is ideally constant, and v is the velocity in the axial direction.

This transducer is relatively delicate but has the advantage of producing an output signal that is relatively large, and therefore requires little or no signal conditioning. Modern electromagnetic velocity transducers are designed using computer-aided design software including finite-element analysis to ensure the quality of dynamic response in the target frequency range.

4.3.4   Accelerometers

Nowadays, however, velocity and displacement are commonly measured using accelerometers, the required parameters being derived by integration. Accelerometers are rigidly fastened to the body undergoing acceleration. For each accelerometer there is a sensing axis aligned with the direction of the acceleration that is intended to be measured. Accelerometers produce an electrical output that is directly proportional to the acceleration in the sensing axis. The output should be low if the acceleration is applied at 90° to the sensing axis; the non-zero output is due to cross-sensitivity.

The piezoelectric device has become almost universally accepted as the transducer to use for all but the most specialised of vibration measurements. The piezoelectric crystals act as both the spring and damper in the accelerometer that is consequently very small and light in weight. The piezoelectric accelerometer is physically much more robust than the velocity transducer and has a far superior frequency range. This has become more important as condition-monitoring techniques involving frequencies well above 1 kHz have been adopted. The construction of a typical piezoelectric accelerometer is illustrated in Figure 4.14.

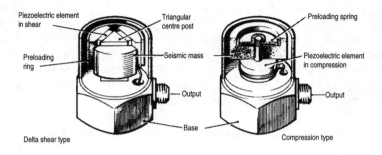

Figure 4.14   A piezoelectric accelerometer [Source: Bruel & Kjaer]

When it is subject to vibration, the seismic mass, which is held against the piezoelectric crystal element, exerts a force upon it. This force is proportional to the acceleration. Under such conditions the piezoelectric crystal, which is usually a polarised ceramic material, generates a proportional electric charge across its faces. The output can then be conditioned using a charge amplifier and either incremental velocity or displacement signals recovered by integration. The device has the obvious advantage of generating its output without an external electrical source being required. As the electrical impedance of a piezoelectric crystal is itself high, the output must be measured with a very high-impedance instrument to avoid loading effects. Integrated circuit piezoelectric devices have become available with a high-impedance charge amplifier resident in the accelerometer encapsulation. In recent years, integrated accelerometer circuits, which include more than one sensing axis, have been developed to permit 2D and 3D measurements.

When using piezoelectric accelerometers it is important to realise that, unlike proximity probes, the natural frequency of the device is designed to be above the usual operating range. A typical frequency response is shown in Figure 4.15. This limits the

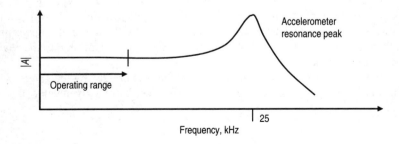

Figure 4.15   Accelerometer response

useful operating range to around 30 per cent of the natural frequency. Also, because the output is low at low frequencies the normal range of application of accelerometers is approximately 1–8 kHz, although small devices may have ranges extending beyond 200 kHz and integrated amplifier allows measurement at nearly DC [9].

There is an extremely wide range of piezoelectric accelerometers available today, from very small devices that will measure shocks of high acceleration (in excess of 10^6 ms^{-2}) to large devices with sensitivities greater than 1 000 pC/ms^{-2}. Highly sensitive devices, on the other hand, have to be physically large so as to accommodate the increased seismic mass required to generate the high output. In all cases, however, care must be taken when mounting accelerometers since they can be easily destroyed through over-tightening. Table 4.3 provides a short summary of the area of application of each of the vibration transducer types discussed.

Table 4.3   *Application of vibration monitoring techniques*

Application	Transducer type
Motor/pump drives	Velocity or acceleration
Motor/fan drives	Displacement or velocity
Motors connected to gear boxes (rolling element bearings)	Acceleration
Motor with oil film bearings	Displacement
Generators steam turbines	Displacement
Overall vibration levels on all of the above	Velocity

4.4   Force and torque measurement

One of the commonest ways of measuring force is to use a strain gauge, a simple device that comprises a long length of resistance wire formed into a zigzag shape and securely bonded to a surface that will alter shape elastically under the action of the force. When the gauge is stressed under the action of the force, the cross-section and length of the wire changes so that its resistance alters. In recent years, wire-type gauges have largely been replaced by metal foil or semiconductor types. Figure 4.16 shows a metal-foil-type strain gauge; the material used is typically the 'advance alloy' made of copper, nickel and manganese. Cutting a foil into the required shape is much easier than forming a piece of wire into the required shape, and this makes the devices cheaper to manufacture.

The input–output relationship of a strain gauge is expressed by the gauge factor, G, which is defined as the relative change in resistance, R, for a given value of strain, ε; that is,

$$G = \frac{\Delta R / R_{\text{nominal}}}{\varepsilon} \tag{4.2}$$

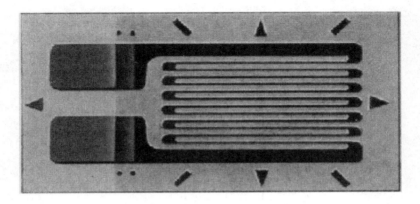

Figure 4.16   A metal foil-type strain gauge

Strain gauges are manufactured to various nominal values of resistance, of which 120, 350 and 1 000 Ω are the most common. The typical maximum change of resistance in a 120 Ω device would be 5 Ω at maximum deflection. The resistance of the gauge is usually measured by a bridge circuit, and the strain (and hence force) is inferred from the bridge output measured. The maximum current that is allowed to flow in a strain gauge is in the region of 5–50 mA, generally small to limit the self-heating effect.

In order to measure forces, strain gauges need to be applied in a particular arrangement, examples of which are shown in Figure 4.17. Within the same stressed

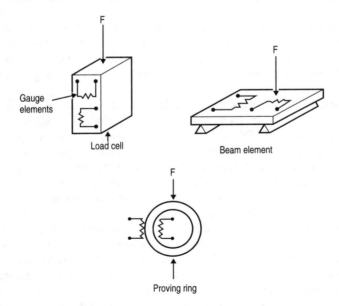

Figure 4.17   Arrangement of strain gauges to measure force

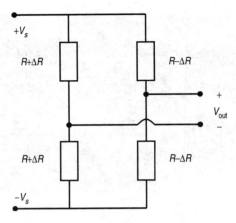

Figure 4.18   Wiring of strain gauges in a Wheatstone bridge

structure, usually two strain gauges are subject to tension while other two are subject to compression. Consequently the resistance of some strain gauges increases while that of others decreases. Figure 4.18 shows a common way of wiring four strain gauges into a Wheatstone bridge where the increased and decreased resistance alternates to maximise the output sensitivity. Note that the voltage supply to the Wheatstone bridge circuit is $\pm V$ rather than between 0 and $2V_s$. This is to avoid common mode voltage in the output that can cause common mode errors in the downstream circuits [10].

Force transducers of this type are routinely used to measure forces ranging from a few newtons to many tonnes force. Care must be taken to either operate them at a constant temperature or properly compensate for the effects due to expansion. Modern strain gauge bridge systems, to measure ΔR, may be self-balancing with automatic read-out.

It is immediately apparent that such devices can be used to measure torques applied to shafts but if the shaft is in motion the additional problem of extracting the signal must be faced. This may be done using instrumentation grade slip rings attached to the shaft, or by suitable noise-free telemetry. Telemetry systems in the electrically noisy environment around electromechanical machinery can be difficult to implement. Alternatively, contact-free inductive (magneto-elastic) or fibre-optic (fibre Bragg grating) torque transducers can be used that work on the change of the material permeability or the reflection of polarised light. Table 4.4 outlines a few such strain measurement methods that, although still relatively expensive, are becoming increasingly popular owing to the computing power available for compensation of temperature and/or non-linearity effects. For instance, Sihler and Miri [11] described a case of using an inductive torque transducer in a synchronous generator system to monitor the occurrence of subsynchronous resonance and this enables the control system to provide damping using active compensation, this subject is dealt with in more detail in Chapter 7.

Table 4.4   Other strain measurement methods

Method	Principle of operation	Remarks
Fluidic load cells	Force applied to a diaphragm causes pressure	Pressure signal must then be converted to an electrical output
Optical encoders	Twisting torque causes displacement across the two ends of a shaft	High resolution needed. Usually applied on long shaft such as in turbine-generators
Fibre-optic	Some fibre-optics exhibit the property to rotate the transmitted light in proportion to the stress applied to the fibre	Careful phase measurement needed. Expensive but highly accurate
Magneto-elastic	The magnetic properties of some ferromagnetic materials depend on the mechanical stress the material is subject to	Expensive but accurate. Special precaution needed to screen devices

In this section, we have simply outlined the commonest means of measuring force, and it is an area that is extremely well established and highly developed. The use of specific techniques for particular applications can be found by searching the large literature associated with force measurement. Some helpful references are Beihoff [12] and Gao *et al.* [13].

4.5   Electrical and magnetic measurement

The basic electrical quantities associated with electromechanical plant are readily measured by tapping into the existing voltage and current transformers that are always installed as part of the protection system. These are standard and therefore need not be considered further here. There can be a requirement, however, to measure the magnetic flux density in or around electrical machines. This is not normally measured for control or protection. For the purpose of condition monitoring, this can be done in one of two ways: either a simple search coil is used, or a Hall-effect device may be more appropriate. These two options are different in nature since the search coil is a passive device whereas the Hall-effect device is not. On the other hand, only the search coil has the capacity for significant energy storage and the risk of sparking; therefore, in areas that may present an explosion hazard, such as on offshore oil rigs or in refineries, their use must conform to current safety standards. These standards are well set out [14]. With the search coil the induced EMF, e, is given by the expression

$$e = \omega BAN \tag{4.3}$$

where ω is the frequency of the normal component of flux density, BA is the effective cross-section of the coil and N the number of turns in the coil. Such devices cannot detect DC fields and at very high frequencies their output may be limited by self-screening effects due to the parasitic capacitance between turns.

Coils can be produced by evaporating copper directly onto surfaces in the appropriate position, or by evaporating them on to insulating materials such as Mylar, which can then be bonded to the appropriate surface. These techniques are extremely useful if the coil is to be used inside an electrical machine, where the coil must not be allowed to move into the airgap where there is a risk of damage to other areas of the machine.

As previously mentioned, if the coil is to be placed in a hazardous area then it is not the output of the coil that is important, it is the possibility of an unwanted input that must be considered. The energy stored in a coil is given by the expression

$$E_e = (1/2)LI^2 \tag{4.4}$$

where L is the inductance of the coil, and I the current flowing in it. The stored energy E_e can be reduced by ensuring that either L is low or I is limited. For a high signal output, L, which is a function of the number of turns, N, should be reasonably high. Therefore one must limit I, usually by a large resistance of the coil. It is then important that the input impedance of the downstream circuit is high. If this is the case, the voltage measured across the terminals of the search coil indicates the EMF and hence the flux density, without much effect of the coil resistance or inductance. A search coil is often made of thin wire so the coil resistance is high; the input impedance of the downstream circuit should be significantly higher. Suitable buffering to the signal conditioning amplifier is sufficient to allow operation in most areas.

The Hall-effect device does not suffer from such disadvantages so long as the required power supply to the device is suitably isolated, but it is by nature only able to provide a measurement of flux density over a very small area. Figure 4.19 shows the basic principle of operation of the Hall-effect element. When a current I flows through the Hall device perpendicular to an applied magnetic field strength B, the electrons are crowded to the front surface due to Lorentz force causing a transverse voltage across the device. The output voltage, V, is related to the applied current, I,

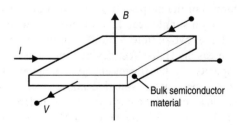

Figure 4.19   The Hall-effect principle

and the field, B, by

$$V = \frac{kIB}{nq} \tag{4.5}$$

where q is the electronic charge, n is the number of charge carriers per unit volume in the semiconductor, k is a constant, and k/nq is the Hall constant of material. In measuring magnetic field the current is established and regulated by a power supply.

Hall-effect devices have the advantage of being able to measure down to DC, and can be made in extremely small sizes. Since Hall effect can be established in less than a nanosecond, the response bandwidth of a Hall-effect transducer can easily be over 100 MHz.

We said at the beginning that current and voltage measurements are usually obtained from current transformers and voltage transformers. This is often the only option we have for large generators in power plants. For small- and medium-size electric machines, for example motors in most variable speed drives, it is often more common to measure both current and voltage using Hall-effect transducers. Figure 4.20 shows a board with three Hall-effect current transducer channels and a board with three Hall-effect voltage transducer channels. The operating principle of the transducer is still that shown in Figure 4.19. Commercial Hall-effect current transducers require voltage supply to establish the current while the magnetic field is set up by the current to be measured. This allows contact-free arrangement as the conductor carrying the current passes through the Hall transducer that is shaped in a ring.

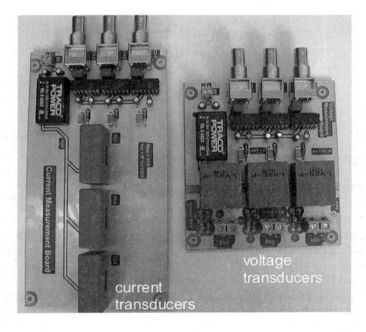

Figure 4.20   Hall-effect current and voltage transducers

Hall-effect transducers are also available for measuring voltage, where the voltage is turned into weak current through a large ohmic resistor, say 100 kΩ. Both current and voltage measurement using Hall-effect transducers requires on-board signal conditioning to set the ratio.

Another current transduction technique used for condition monitoring of electrical machines is the Rogowski coil whose operating principle is similar to that of a search coil. However, a Rogowski coil is specifically arranged to provide features that are desirable in some applications. As shown in Figure 4.21, a Rogowski coil is an air-cored toroidal coil that can be placed around a conductor that may carry current. An EMF is induced by the alternating magnetic field. Since there is no saturable magnetic component in the coil, the total magnetic field is strictly proportional to the current that excites it and the net effect of any current external to the toroidal coil is always zero. To complete the Rogowski coil current transducer, the induced EMF is electronically integrated so that the output from the integrator, properly initialised, is a voltage that accurately reproduces the current waveform. Rogowski coils are often used to capture the high-frequency spikes in the neutral current that may be used to interpret internal faults or partial discharge in an electrical machine [15]. Conventional CTs are not suitable for this purpose due to more limited bandwidth and possible core saturation.

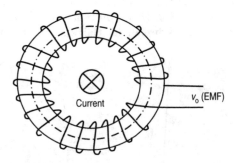

Figure 4.21   Illustration of a Rogowski coil measuring current in a busbar

4.6   Wear and debris measurement

In many electrical machines, we can derive crucial condition information from fluid lubricant and coolant. For example, bearing damage accounts for at least 50 per cent of rotating electrical machine failures (see Chapter 3 and Reference 16). As the bearing wears, we would naturally expect to detect debris in the lubricant. Similarly, when internal faults such as partial discharges occur in the machine, we would expect that the coolant would be contaminated. Chemical methods of monitoring the wear and contamination will be discussed more fully in Chapter 6. For completeness, we will here briefly describe the physical techniques of debris measurement in line with the transduction task in monitoring. The most common method of doing this is to use a so-called debris sensitive detector. Such a detector depends on the lubricating fluid being continually passed through a device that is sensitive to the presence of particular material. This is commonly achieved using either an electrical transducer to measure electrical changes in inductance, capacitance, or conductivity, or optically by measuring changes in turbidity of the lubricant.

The principle of operation of the electrical techniques is essentially the same. The lubricant and debris pass through a small chamber that can alternatively be part of a conductive circuit, the dielectric in a capacitor, or part of a magnetic circuit, in order to measure changes in conductivity, capacitance or inductance, respectively. AC power supply is usually needed when detecting the change of capacitance or inductance. All of these devices can give good indications of general levels of wear, and dramatic indications of the occasional large pieces of debris. Such transducers are both complex and expensive, and require careful and regular attention. Figure 4.22 shows a scheme of measuring the change of inductance, which is most widely used in practice. An inductive coil is placed around the tube carrying the fluid that may contain metallic debris. The inductance of the coil is detected using a high-frequency oscillating circuit. Ferromagnetic debris in the oil increases the coil inductance and hence decreases the resonant oscillating frequency of an LC circuit. If the metallic debris is non-ferromagnetic, the high-frequency inductance of the coil tends to reduce due to eddy current induced in the debris [17].

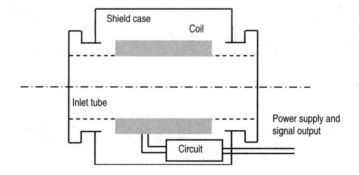

Figure 4.22   Structure of an inductive debris detector

Optical monitors generally operate by sensing either a loss of transmission of light through a test cell, or by detecting light scatter from the particular matter (known as the Tyndall effect), as shown in Figure 4.23.

Maintenance of optical systems is minimal, and they may be used on a wide variety of fluids. They are subject to spurious output, however, if the lubricant becomes aerated to any significant degree. Being more suitable for detecting fine particles in the fluid, they are however unable to differentiate between harmful and non-harmful particulate matters. Optical and electrical detectors are often used in combination in condition-monitoring applications.

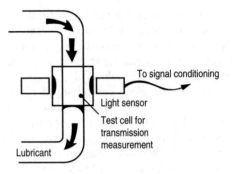

Figure 4.23   Arrangement for an optical debris sensor

4.7   Signal conditioning

Signals from sensing devices will often require some type of modification before they are transferred for data acquisition. This requires local signal conditioning circuits that usually employ analogue devices. The modification may be amplification or averaging, for example. In noisy environments, it is often necessary to filter the outputs of the transducers. Figure 4.24 shows two typical signal conditioning circuits using operational amplifiers and their input–output relationships under ideal conditions. Their operation is explained thoroughly in many textbooks. Figure 4.25 shows a buffer amplifier circuit that can avoid the effect of the resistance of the leads. This allows transmission of analogue signals over distances up to 100 m using screened cables.

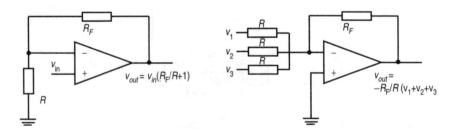

Figure 4.24   Two commonly used amplifier circuits

We saw in previous sections that many transducers output a differential voltage signal by using a Wheatstone bridge whose loading capability is very small; that is, the current drawn from the bridge should ideally be zero. Furthermore, we often would like to turn the differential voltage into a voltage signal referenced to the common ground with the rest to the downstream circuit. This can be achieved by an instrumentation amplifier, as shown in Figure 4.26, for an RTD temperature transducer. The entire instrumentation amplifier circuit is usually made in a single chip and value R_G can be tuned externally to adjust the overall gain of amplification [10].

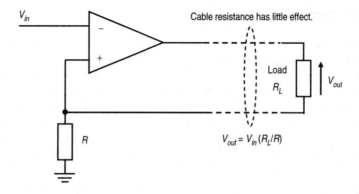

Figure 4.25   Buffer amplifier circuit for signal transmission

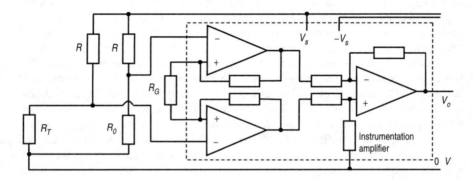

Figure 4.26   Instrumentation amplifier

More generally, great care must be taken with regard to the cabling and termination of transducers and equipment to reduce noise problems [18]. Cables should be of high quality and routed through unexposed areas wherever possible, so as to avoid the possibility of accidental damage. If many channels coexist in the same transmission corridor, care must be taken to avoid common impedance coupling between channels by segregating power from signal cables and ensuring that cables cross at right angles where Electromagnetic Compatibility (EMC) isolation is important. In most cases twin-screened twisted pairs should be used to couple a transducer to the primary data acquisition unit to avoid induction from low frequency fields <500 Hz. Screened coaxial cables are suitable to screen from high frequency fields >100 kHz. Optical fibre cabling is now robust and therefore a frequently used alternative, particularly when high noise immunity is essential. These issues are particularly important with condition-monitoring electrical machines supplied from harmonic rich converters. Communications between the primary data acquisition unit and the processing system should also be established via a high-integrity low-noise link to avoid any chance of data corruption or interruption.

4.8   Data acquisition

The precise nature of the data acquisition techniques used is usually determined by the subsequent algebraic manipulations that will be performed on the datasets produced by the transduction process. It is not really possible, therefore, to separate these two tasks fully in most cases. What is of paramount importance, however, irrespective of the complexity of the monitoring system, is the fidelity of the information received by the processing unit. Transmitted or recorded data must be sufficiently noise-free to comply with the demands of the monitoring system and must be wholly consistent. If this is not the case then the whole dataset, perhaps spanning several months, may be effectively corrupted and therefore useless. In complex systems handling many inputs, either continuously or on a sampled basis, it is usual to have the processing system remote from the plant. In such cases some degree of local data conversion may be advisable. For example the signals taken from a group of machines in adjacent locations may be routed to a nearby collection point that digitises the incoming signals and identifies them for onward transmission to the central system. In noisy environments it may even be desirable to digitise signals at the collection point, and then forward them to a gathering point.

Figure 4.27 illustrates a preferred structure for a fully automatic system. The basic data acquisition system shown earlier is broken down into three concatenated functions: (1) multiplexing; (2) sampling and holding; and (3) analogue-to-digital conversion. An analogue multiplexer directs different signal channels for downstream processing and allows the channel identity to be passed through. The use of a multiplexer is essential if a large number of channels are to be monitored, and can be appropriate even for small numbers of channels because it allows the use of a single high-quality analogue-to-digital conversion, rather than recourse to many

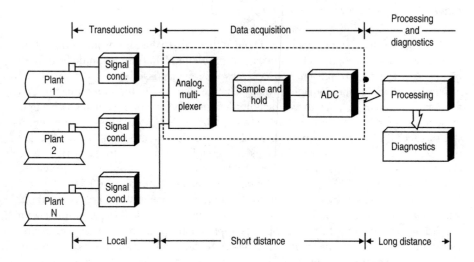

Figure 4.27   General arrangement of data acquisition

lower-grade devices. Commercial multiplexers can now accommodate many channels. An analogue-to-digital converter turns an analogue voltage signal into a digital number that can be recognised by the computer according a defined comparison scheme or scaling factor. During conversion time, the input analogue signal is sampled and the sample is held constant. Resolution of the analogue-to-digital converter, which affects the sensitivity of the condition-monitoring system, depends on the number of bits. Since condition monitoring aims to capture weak signals at early stages of impeding faults, high sensitivity is desirable. A 12-bit analogue-to-digital converter has a resolution of 1 in 4 096 and typically completes conversion in less than 0.01 μs. The switching rate of the multiplexer can therefore be in the range of 10 MHz, adequately high to avoid any significant time lag between the multiplexed channels.

General-purpose data acquisition units, designed in line with the generic structure described earlier, are now commercially available and have naturally been used for condition-monitoring purposes. For example Figure 4.28 shows an 8-channel, 14-bit data acquisition board that can interface with a standard PC and operate in a specific software environment. This is a suitable choice provided that the use of a dedicated PC can be arranged. Other options such as LabVIEW are also available, which normally provide output channels to implement some simple control actions.

Figure 4.28   A commercial data acquisition device to interface to a PC [Source: Adept]

In data acquisition, it is important to determine how often the signal is sampled and how many samples are kept for further processing. This is an area where mistakes can be easily made and we will examine this issue in the next chapter after some signal processing techniques have been explained. Here we just mention that the Shannon sampling theorem must be observed. The theorem states that the sampled data can be used to faithfully reproduce a time-varying signal provided that the sampling rate is at least twice the highest frequency present in the signal. This minimum sampling rate is called Nyquist frequency. Given a sampling rate, the total number of samples kept for signal processing implies the duration during that the signal is sampled. This duration determines the frequency resolution that a spectrum analysis performed on the data can give. For example, if a signal is sampled for 1 s, the frequency resolution is 1 Hz.

4.9   Conclusion

This chapter has shown the wide range of instrumentation and data acquisition techniques now available to monitor electrical machines. Modern techniques and methods have been shown and in the following chapter we show how the signals from these sensors can be processed to provide condition-monitoring information.

4.10   References

1. Morris A.S. *Measurement & Instrumentation Principles*. Oxford: Butterworth-Heinemann; 2001.
2. Horowitz P. and Hill W. *The Art of Electronics*. Cambridge: Cambridge University Press; 1989.
3. Boiarski A.A., Pilate, G., Fink T. and Nilsson N. Temperature measurements in power plant equipment using distributed fiber-optic sensing. *IEEE Transactions on Power Delivery* 1995; **PD-10**: 1771–78.
4. Miyazaki A., Takinami N., Kobayashi S., Nishima H. and Nakura Y. Long-distance 275 kV GIL monitoring system using fiber-optic technology. *IEEE Transactions on Power Delivery* 2003; **PD-18**: 1545–53.
5. Rogers A.J. Distributed optical-fibre sensors. *Journal of Physics D: Applied Physics* 1986; **19**: 2237–66.
6. Kher S., Gurram S., Saxena M.K.and Nathan T.P.S. Development of distributed fibre optic sensor with sub-metre resolution. *Current Science* 2004; **86**: 1202–204.
7. Nandi S., Toliyat H.A. and Xiaodong L. Condition monitoring and fault diagnosis of electrical motors – a review. *IEEE Transactions on Energy Conversion* 2005; **EC-20**: 719–29.
8. Iwaki, H., Yamakawa H. and Mita A. FBG-based displacement and strain sensors for health monitoring of smart structures. Presented at the 5th International Conference on Motion and Vibration Control, Sydney, Australia; 2000.
9. Bruel & Kjaer. *Machine Health Vibration*. Naerum: Bruel & Kjaer; 1984.
10. Bogart Jr T.F., Beasley J.S. and Rico G. *Electronic Devices and Circuits*. New Jersey: Pearson Educational; 2004.
11. Sihler C. and Miri A.M. A stabilizer for oscillating torques in synchronous machines. *IEEE Transactions on Industry Applications* 2005; **41**: 748–55.
12. Beihoff B. A survey of torque transduction methodologies for industrial applications. Presented at the IEEE Pulp and Paper Industry Technical Conference, Birmingham, AL; 1999.

13. Gao G., Steinhauser M., Kavanaugh R. and Chen W. Using fiber-optic sensors to measure strain in motor stator end windings during operation. Presented at the Electrical Insulation Conference and Electrical Manufacturing & Coil Winding Conference, Cincinnati; 1999.
14. Garside R. *Intrinsically Safe Instrumentation Safety Technology*. London: Feltham; 1982.
15. Xiao C., Zhao L., Asada T., Odendaal W.G. and van Wyk J.D. An overview of integratable current sensor technologies. *Proceedings of the 38th IEEE IAS Annual Meeting*; Salt Lake City, 12–16 Oct 2003; **2**: pp. 1251–8.
16. Durocher D.B. and Feldmeier G.R. Predictive versus preventive maintenance. *IEEE Industry Applications Magazine* 2004; September/October: 12–21.
17. Whittington H.W., Flynn B.W. and Mills G.H. (1992). An on-line wear debris monitor. *Measurement Science Technology* 1992; **3**: 656–61.
18. Williams T. *EMC for Product Designers*. Oxford: Newnes; 1996.

Chapter 5
Signal processing requirements

5.1 Introduction

Broadly speaking, the processing task is that part of the monitoring activity where data, which has been collected and suitably formatted, is operated upon or otherwise transformed so that a diagnosis of plant condition can readily be made. As we have mentioned in the previous chapter, it is here that the most significant scope for automation exists. Indeed, to perform many of the data processing functions now commonly used in condition monitoring, considerable computational power is required. Processing may be done on- or off-line and this choice will predominantly depend upon whether the monitoring system is one that operates on a continuous basis or not. It also depends on the understanding regarding how quickly the monitored faults can develop. Usually, condition monitoring in a continuously operating plant or those where faults that can develop quickly, should allow the obtained data to be processed on-line. However the distinction cannot be clearly drawn. Modern data acquisition units now permit data to be obtained from a running plant and analysed at a different location without any significant time delay.

In the past, dedicated signal analysers were applied to analogue voltage signals from the transduction process to obtain the spectra that could indicate the occurrence and level of signal components at particular frequencies. Their use has now almost been completely replaced by general computer algorithms acting on data acquired in a digital format.

Although commercial computer software packages are available for digital signal processing applications and can be used for condition monitoring of electrical machines, it is important to fully appreciate the characteristics of the different algorithms in order to avoid unreliable results being obtained, leading to misinterpretation of the plant condition. Requirements of such algorithms and their limitations must be clearly understood. For such a reason, this chapter describes in greater detail some of the more important numerical techniques used for signal processing in condition monitoring.

Perhaps the simplest form of signal processing is one in which the magnitude of the raw incoming signal is examined on a regular basis, as a function of time. In fact this is essentially the basis of all visual inspection techniques or trend analysis, which involves active collection of data by personnel. The processing in such cases may consist of a comparison of the current record with the previous value or with some preset or predetermined threshold. This process is simple to automate, even when many hundreds of inputs are being monitored. The processor is simply required to associate the incoming reading with a particular item of plant. If the incoming data is to be trended, to examine the lead up to an atypical event, then a data storage medium is essential. Magnetic disk or optical storage is the commonest method. Semiconductor memory may also be appropriate provided that it can be protected against instrument power failure with the use of a battery back-up system. It is easy, when monitoring many inputs, to accumulate exceedingly large volumes of data, so it is desirable to automatically refresh the data storage so that after a given period of time the data storage elements either transfer their contents to a bulk storage media, or are overwritten by the incoming data stream. When a plant malfunction is detected, a common practice is to annunciate the event and discontinue data storage on that channel. In this way the diary of events leading up to the malfunction is preserved for subsequent examination if required.

The earlier example of magnitude detection implies that the plant condition information is seeded in the change of signals obtained from the plant. Indeed, the variation of signals usually tells us the change of condition. Variation of a signal in the time domain can be more saliently expressed as components in the frequency domain. Therefore spectral analysis is naturally a very common technique of signal processing used in condition monitoring. Spectral analysis is effective when applied to steady-state periodic signals, which is usually the case with monitoring machine faults that gradually develop. In recent years, higher-order spectral analysis has been used to make use of phase information and improve the signal-to-noise ratio. We will describe first- and higher-order spectral analysis techniques in this chapter. Sometimes a condition-monitoring technique may also be applied to transient and non-stationary signals. Wavelet transform techniques that have more recently been developed will then be described for this objective. Furthermore, correlation analysis, which is a powerful tool in revealing the buried relationships between signals and hence identifying the causes and consequences of fault, will also be described, together with some other basic signal processing techniques that are frequently used in condition monitoring.

5.2   Spectral analysis

Spectral analysis is the name given to describe methods that transform time signals into the frequency domain. The spectral representation of a time signal is therefore a collection of components in the frequency domain, each with a specific frequency, amplitude and phase angle. The transformation is achieved using the techniques of Fourier analysis whereby any periodic signal, g, of period T, has

$$g(t) = g(t + T) \tag{5.1}$$

This can be represented by equally spaced frequency components $G(f)$:

$$G(f_k) = \frac{1}{T} \int_{-T/2}^{T/2} g(t) e^{-j2\pi f_k t} dt, k = 0, \pm 1, \pm 2 \tag{5.2}$$

where subscript k indicates the kth harmonic of the fundamental frequency $f = 1/T$. The harmonic represents a sine wave at the harmonic frequency whose amplitude and phase angle can be determined from (5.2). The original time domain signal can be reconstructed by summing all the harmonic components as shown in (5.3).

$$g(t) = \sum_{k=-\infty}^{\infty} G(f_k)e^{j2\pi f_k t} \tag{5.3}$$

We see that a continuous periodic time function can therefore be represented as a discrete series in the frequency domain. The advantages are immediately apparent; our continuous input can be approximately represented to the required accuracy by a finite and often small set of numbers. As we will illustrate in later chapters, many faults give rise to harmonic components that are normally non-existent or of insignificant amplitude. Spectral analysis compacts the data significantly at the cost of computation time, and allows trends of change to be more easily recognised.

Because it is often necessary to digitise the transducer signals for onward transmission to the processing unit in a monitoring system, the incoming data is also sampled in time. That is, it is represented as a series of discrete values at equally spaced instants in time, in a similar fashion to the frequency domain representation of a continuous time signal. Under these conditions the equivalent statement of (5.2) becomes

$$G(f_k) = \frac{1}{N} \sum_{n=0}^{N-1} g(t_n)e^{-j2\pi nk/N} \tag{5.4}$$

and the corresponding inverse transform is

$$g(t_n) = \sum_{k=0}^{N-1} G(f_k)e^{j2\pi nk/N} \tag{5.5}$$

In these expressions we see that the frequency is effectively sampled at the discrete frequencies f_k while the time signal is sampled at instant t_n. We therefore have a means of representing a discrete time function by a set of discrete values in the frequency domain. This transformation is known as the discrete Fourier transform. In practice this transformation is carried out using the fast Fourier transform technique, which is an extremely efficient way of achieving a discrete Fourier transform.

When sampling a continuous signal, the total sampling length (the observation interval) must be taken into consideration. The entire sampled data is automatically treated as one period in spectral analysis. If the signal is not strictly periodic or the sampled data does not precisely represent an actual period, then harmonic distortion may occur because the periodic waveform created by the sampling process may have sharp discontinuities at the boundaries. This effect may be minimised by windowing the data so that the ends of the data block are smoothly tapered. Common window functions include Bartlett, Hanning, Hamming and Blackman windows [1]. The original sampled data is multiplied with the window function before fast Fourier transform is applied to generate the frequency domain spectrum, as shown in Figure 5.1. What

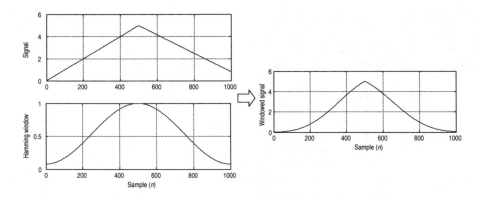

Figure 5.1   Illustration of window function

we obtain is not a true spectral representation of the original time domain signal, but the effect of boundary discontinuities is attenuated. A way to view the windowing technique is that the finite sampling process itself has already windowed the true waveform with a rectangular window. Applying an additional window will give a more desirable result that minimises frequency domain distortion.

Another way often used in practice is to increase the length of the sampled data block, by increasing the observation interval. This lowers the fundamental frequency assumed in the spectral analysis. For a given sampling rate, the observation interval is proportional to the number of samples taken for the fast Fourier transform. With increased observation interval, the harmonic distortion due to the discontinuity at boundaries tends to concentrate in the low-frequency range; their contribution in the high-frequency range is reduced. It must be understood that the finite observation interval always results in a fundamental limit on the frequency resolution. Lengthening the observation interval increases the frequency resolution in the resultant spectrum.

Since fast Fourier transform only provides components at discrete harmonic frequencies, which are integer multiples of the fundamental frequency, there is always the risk that the target component for condition monitoring is at a frequency that is between two integer harmonic frequencies. This means that not all frequencies can be seen by the fast Fourier transform, due to the so-called 'picket fence' effect, disregarding the length of the observation interval. In such a case, the energy of the components unseen by the fast Fourier transform slips through the picket fence and leaks into the harmonics that are seen by the fast Fourier transform. Several algorithms, including polynomial estimations, have been developed to detect the target component from the harmonics that have been identified by fast Fourier transform. For example Figure 5.2 shows how the energy of harmonics of unity amplitude of a machine current signal at 59.83, 59.85 or 59.87 Hz could leak to components at 59.6, 59.7, 59.8, 59.9, 60 and 60.1 Hz that are at the fast Fourier transform output frequencies, and this depends on the sampling scheme used. Details of the algorithms to estimate the components at intermediate frequencies, as a reverse process of what

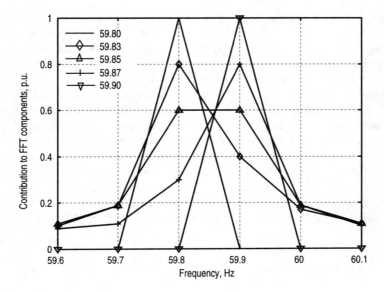

Figure 5.2   Illustration of picket fence leakage

is shown here, can be found in Reference 2. In condition monitoring of rotating electrical machines, the target frequency component in a signal is usually dependent on the speed of the machine that may vary continuously depending on the load condition. As a result, the target frequency is unlikely to be a multiple integer of the fundamental frequency determined by the length of the observation interval.

A final point that we would like to highlight here is the risk of aliasing that can occur in the spectral analysis. More and more electrical machines nowadays are connected to power electronic converters, which generate harmonics. High-frequency harmonics exist in both machine current and vibration signals that may be used for condition monitoring. Figure 5.3 illustrates the cause of aliasing. Suppose the original continuous time signal has a spectrum shown in Figure 5.3(a). The sampled discrete time data then will have a spectrum shown in Figure 5.3(b) where the original spectrum is duplicated according to integer multiples of the sampling frequency [3]. If the

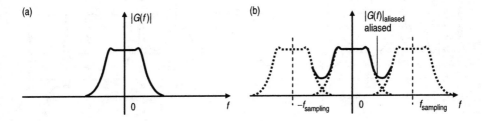

Figure 5.3   Occurrence of aliasing. (a) True spectrum. (b) Aliased spectrum.

sampling frequency is lower than the Nyquist frequency that is twice of the highest frequency in the signal, as shown in Figure 5.3, then aliasing occurs. In this case, some harmonic components in the spectrum obtained by performing fast Fourier transform on the sampled signal are due to aliasing or sampling effects rather than real components of the original continuous time signal.

Consider an example. Figure 5.4(a) shows the stator current of an inductor motor fed by a PWM (pulse-width modulation) voltage source inverter with a symmetrical PWM switching frequency of $f_{sw} = 2.5$ kHz. The current is then sampled, also at a rate of 2.5 kHz, and the result is shown in Figure 5.4(b). The sampling rate may be considered natural because it coincides with the internal clock of the controller that generates the PWM pulses. Suppose that the output fundamental frequency of the inverter is $f_o = 10$ Hz. The actual current then contains a fundamental component and harmonics at $f_{sw} \pm 2kf_o$ where $k = 1, 2, 3 \ldots$, giving harmonics at 2 480 Hz, and 2 520 Hz, due to the modulation process in the inverter [4]. A section of the

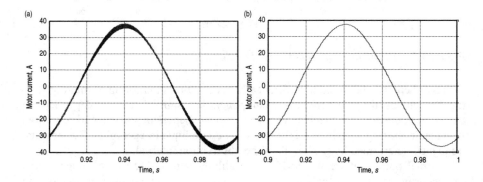

Figure 5.4   *Inverter fed motor current. (a) Inverter fed motor current. (b) Sampled motor current.*

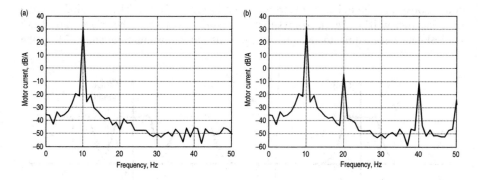

Figure 5.5   *Effect of sampling on inverter fed motor current spectrum. (a) True spectrum of motor current. (b) Spectrum of sampled motor current.*

true spectrum is shown in Figure 5.5(a). The corresponding spectrum of the sampled current in Figure 5.4(b) is shown in Figure 5.5(b). The 2 480 Hz and 2 520 Hz components in the original spectrum, after being shifted downwards by 2.5 kHz, have contributed to the 20 Hz component in the aliased spectrum that may be misinterpreted in condition monitoring and therefore should be taken care of. The 40 Hz component is caused by the same mechanism. The effect of sampling also depends on the instant of a sample is taken in a PWM switching cycle [5].

If the sampling rate cannot be further increased, then a common practice is to physically filter the signal and limit its frequency bandwidth before it is sampled. This anti-aliasing filter can prevent unrealistic components being introduced to the resultant spectrum. In general, the filter bandwidth and the sampling rate afterwards should be determined after knowing the target frequencies that are to be extracted for the purpose of condition monitoring.

5.3   High-order spectral analysis

High-order spectral analysis is a relatively new technique used for condition monitoring. The first-order spectral analysis as described in the previous section makes little use of the phase information of the Fourier harmonic components. This can be taken into account in higher-order spectral analysis and the signal-to-noise ratio can be consequently improved [6].

For first-order harmonic components defined by (5.1), a second-order power spectral density at frequency f_k can be defined as

$$P(f_k) = G(f_k)G^*(f_k) \tag{5.6}$$

where $G^*(f_k)$ is the complex conjugate of $G(f_k)$. Extending this definition to third- and fourth-order measures gives birth to the bispectrum $B(f_1, f_2)$ and trispectrum $T(f_1, f_2, f_3)$ as shown below.

$$B(f_1, f_2) = G(f_1)G(f_2)G^*(f_1 + f_2) \tag{5.7}$$

$$T(f_1, f_2, f_3) = G(f_1)G(f_2)G(f_3)G^*(f_1 + f_2 + f_3) \tag{5.8}$$

From (5.7) and (5.8), we may see that, unlike the power spectral density $P(f_k)$, the bispectrum and trispectrum are functions of more than one frequency index and, further, we may also see that they are complex quantities; that is, they contain both magnitude and phase information about the original sampled time signal. In contrast, $P(f_k)$ contains only real values.

We can appreciate the value or potential of such high-order spectral analysis by looking into what happens during some typical faults that cause electromagnetic disturbances in the machine. For example, a broken rotor bar fault in an induction motor will usually give rise to harmonics at two frequencies f_1 and f_2 in the airgap flux density, with phase angles ϕ_{f1} and ϕ_{f2}. A vibration signal is likely to be related to the flux according to a quadratic function [7]. As a result, the vibration signal that can be acquired through an accelerometer contains harmonics at frequencies f_1, f_2 and $f_1 + f_2$ due to the quadratically non-linear relationship. Their phase angles will be

ϕ_{f1}, ϕ_{f2} and $\phi_{f1} + {}_{f2} = \phi_{f1} + \phi_{f2}$, showing a quadratic phase coupling that can be easily indicated by a peak in the bispectrum of the vibration signal at the bifrequency, $B(f_1, f_2)$, where the associated biphase $\phi(f_1, f_2)$ tends to zero. Faults of other types that cause flux components at different frequencies can be similarly detected provided that the target frequencies are known. Higher-order spectral analysis can be used when multiple faults exist in the machine so that more signature frequencies can be identified with a high signal-to-noise ratio, by exploiting the property of phase coupling. Examples of using high-order spectral analysis for condition monitoring of induction machines are shown in Reference 6.

5.4   Correlation analysis

This is a time domain technique. Correlation between two signals, in the time domain, is a process mathematically very similar to that of convolution. The auto-correlation function provides a measure of the similarity between a waveform and a time-shifted version of itself, while the cross correlation function refers to two different time functions.

The auto-correlation function of a time signal $f(t)$ can be written as

$$R_{ff}(\tau) = \int_{-\infty}^{\infty} f(t - \tau)f(t)dt \tag{5.9}$$

The function $f(t - \tau)$ is delayed version of $f(t)$, by a time τ. Essentially the process may be thought of as one signal searching through another to find similarities.

If $R_{ff}(\tau)$ is plotted against τ, then the result is a correlogram. If the signal tends to repeat, the correlogram plot will give a peak indication when τ is around the time that it takes the signal to show some repetition. Therefore auto-correlation is a powerful tool to identify the repeating features that can be hidden in a signal mixed with noises and disturbances. Figure 5.6 illustrates the principle. The time signals are shown in Figure 5.6(a) and 5.6(b), and the resulting correlogram in 5.6(c). It is apparent that signals exhibiting a periodicity, which may be difficult to extract from a background of noise, can be identified using this technique.

The cross-correlation function of two different signals $f(t)$ and $h(t)$ may be written as

$$R_{fh}(\tau) = \int_{-\infty}^{\infty} f(t - \tau)h(t)dt \tag{5.10}$$

The similarity between correlation and convolution is now clearly seen. For example the convolution of the two time functions $f(t)$ and $h(t)$ would be expressed as

$$g(t) = \int_{-\infty}^{\infty} f(t - \tau)h(\tau)d\tau \tag{5.11}$$

The only difference is that one of the time functions is effectively reversed.

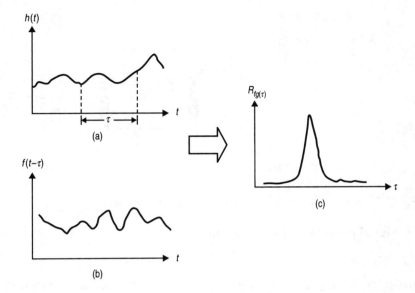

Figure 5.6   The use of correlation functions

Like the auto-correlation function, the cross-correlation function can also be used to recover both the amplitude and phase of signals lost in a noisy background. If two signals are inherently related but are phase shifted due to transmission delay in the system or modulated by physical processes in the system, cross-correlation function can reveal the similarity between these signals, which are then confirmed to be due to the same cause. Because of such a property, correlation functions are particularly suitable to monitor faults like bearing degradation and partial discharge activity on worn transformers, including their on-load tap changers, which have intermittent effects during operation [8,9]. Correlation functions provide a means of relating the condition-monitoring signatures to the causes of faults. Correlation functions can also be used to exclude the effects of factors that may not have direct implication on the plant condition, for example known variations of load or changes of ambient temperature.

If we redefine (5.9) and (5.10) in a slightly different way, that is let the cross-correlation function between two signals $f(t)$ and $h(t)$ be

$$R_{fh}(\tau) = \lim_{T \to \infty} \frac{1}{T} \int_0^T f(t - \tau)h(t)dt \qquad (5.12)$$

it is easy to see how the correlation process can be realised in a numerical algorithm. Figure 5.7 illustrates the activity schematically; the time averaging operation required will be described in the next section.

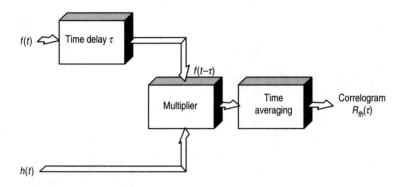

Figure 5.7   Implementing the measurement of correlograms

5.5   Signal processing for vibration

5.5.1   General

Vibration monitoring requires a number of specialised techniques and two shall be briefly mentioned here. Although they are beginning to establish themselves as powerful diagnostic tools, their acceptance (and subsequently their application) is at the moment limited primarily to the diagnosis of faults in gearboxes.

5.5.2   Cepstrum analysis

Mathematically the cepstrum, $C(\tau)$, of a time function is described as the inverse Fourier transform of the logarithm of the power spectrum of the function; that is, if we define the power spectrum, $P_g(f)$, of a time function $g(t)$ as

$$P_g(f) = F\{g(t)\}^2 \tag{5.13}$$

then the corresponding cepstrum is:

$$C(\tau) = F^{-1}\{\log_e P_g(f)\} \tag{5.14}$$

where F and F^{-1} represent the forward and inverse Fourier transforms described in Section 5.2. Therefore $P_g(f)$ is the power spectrum of the time signal $g(t)$, which is also defined in (5.6). The dimension of the parameter τ in (5.14) is time and is introduced by the inverse Fourier transform; hence we can display the magnitude of the cepstrum with respect to time intervals in the same way as the spectrum can be illustrated with respect to frequency. The logarithm is employed so that the harmonic components in $P_g(f)$ that have relatively low amplitudes are also taken into account; their existence is highlighted. The inverse Fourier transform in (5.14) tries to reconstruct the time signal but only highlights the time instants when considerable activities are occurring. From this, the repetition of the activities can be identified, which is demonstrated next [10].

The use of the cepstrum has found favour in examining the behaviour of gearboxes because such items of equipment tend to produce many families of side bands in their vibration spectra, due to the variety of meshing frequencies and shaft speeds that may be present. Figure 5.8(a) shows the power spectrum of a gearbox vibration signal, containing harmonic side bands that are spaced according to a pattern. We hope to identify such a pattern in order to derive information about the condition of the gearbox and the cause of fault, if present. Figure 5.8(b) shows the cepstrum calculated from the power spectrum using (5.14). The horizontal axis here is practically time and the spikes correspond to the instants when significant activities are present in the original time signal. It is clear that the activities repeat predominantly according to two time periods, marked as A and B respectively in Figure 5.8(b). By measuring the periods (8.083 ms and 19.6 ms), we know that the fundamental repetition frequencies are 123.75 Hz and 51 Hz. We can then further infer that the power spectrum of

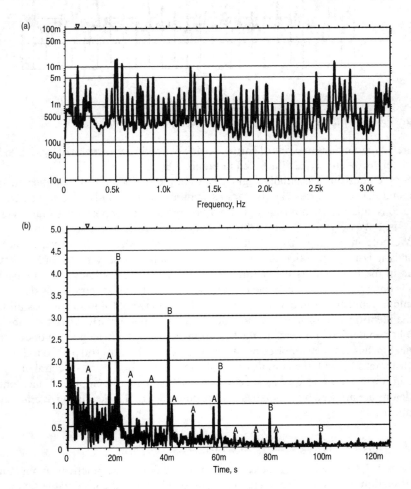

Figure 5.8   Vibration spectra and associated cepstra. (a) Power spectrum of a vibration signal. (b) Cepstrum of the gearbox vibration signal.

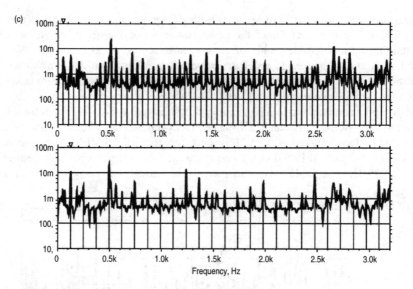

Figure 5.8   Continued (c) Power spectrum of a vibration signal.

Figure 5.8(a) contains two sets of harmonic side bands that are separated by 123.75 Hz and 51 Hz respectively, as illustrated in Figure 5.8(c).

So the cepstrum essentially highlights multiple periodicity in complicated signals, and hence identifies clearly various families of side bands. The identification of various side bands in a rich signal may be practically impossible using spectral analysis, but as Figure 5.8 shows the cepstrum easily picks them out.

We note in passing that although the horizontal scales of the cepstrum are in seconds it is usual practice to refer to the horizontal quantity as 'frequency', the peaks in the cepstrum as the 'harmonics' and the process is known as 'filtering'. This is done to firmly identify the methodology with that of spectral analysis. If this technique can identify side bands with ease, then it may hold significant possibilities for the identification of faults in induction machines, particularly when they are fed from harmonic rich inverters. To the best of the authors' knowledge no work on this application has yet been reported but it may yet prove to be a fruitful path to follow. Its disadvantages lie with the complexity of the technique. It is a post-spectral analysis tool, in much the same way as spectral analysis is generally employed once one's suspicions are aroused by anomalies in time domain signals taken using overall level monitoring.

5.5.3   Time averaging and trend analysis

In Figure 5.7, the need for signal averaging is noted during the practical computation of correlation functions. The averaging technique has found considerable favour in its own right for the detection of faults in gearboxes and rolling element bearings. It achieves this by simply averaging a large number of samples, taken successively from the same transducer or an upper stream numerical process, with each sample carefully timed to the same period. With rotating plant the rotational period of the element under inspection is usually chosen. In this way noise generated elsewhere in the system is effectively smoothed out, and the signals due to faults that exhibit a cyclic pattern over the chosen period, enhanced.

Averaging can be implemented using one of the many transient signal capture techniques. For example if it is necessary to take the average of 1 000 records, say, then each record must be captured and then stored for a simple average to be obtained on a point-by-point basis. This is an inefficient technique since it requires a large amount of digital storage. It is possible to reduce this requirement significantly by approximation and defining the average of a sampled quantity, x_n, as

$$\bar{x}_n = \bar{x}_{n-1} + \frac{x_n - \bar{x}_{n-1}}{n} \tag{5.15}$$

where n is the record number, x_n is the value for the nth record, and $\bar{x}_n$ and $\bar{x}_{n-1}$ are the average values up to the nth and $(n-1)^{\text{th}}$ records respectively. Using this technique it can be seen that only two records need to be stored at any given time, yet the truly moving average is still calculable. The corresponding saving of digital memory is obvious, but it has been gained at the expense of requiring more rapid processing hardware, particularly if the period of each record is short. This problem is not always significant, particularly as fast processors are becoming more easily available in embedded systems. Time averaging is seldom applied to a signal that changes very quickly with time. For example, it should not be used for time domain voltage and current signals. Time averaging is usually applied to a signal that indicates the level of a signature in the frequency domain, for example the rms value of the voltage or current. Time averaging helps to avoid the effect of random background noise in the signal. It is nearly always possible to choose records with suitable gaps between them. It is however of paramount importance to ensure that each record is properly synchronised. In addition to time averaging that filters out the random noise in the measurement, it is often necessary to perform some trend analysis to capture the change of a condition-monitoring signature in a relatively long period of time. Auto-regression models in time series analysis [11,12] can be applied for the purpose on monitoring developing fault.

The usefulness for time averaging techniques in monitoring electrical machines is somewhat limited although it is effectively the technique described by Tavner *et al.* [13] for detecting rotor cage faults by speed variation. It may also prove helpful for drive systems where the effect of a gearbox, excluding the input shaft and bearings, can be removed using this technique.

5.6 Wavelet analysis

We do not attempt to present even an overview of the rich subject of wavelet transform; interested readers are referred to Flordin [14] for example. But there is increasing interest in developing condition-monitoring techniques based on signal processing methods using wavelet transform. This is primarily because of the limitations of conventional spectral analysis based on Fourier series. For example, spectral analysis assumes that a signal is periodic and the harmonic components are obtained as average over the entire observation interval; there is no information about local variation of a signal at certain frequency during a particular short period of time. Faults of an intermittent nature, such as mechanical faults on the drive train of an electric machine system, cause intermittent, non-stationary, oscillatory signals (e.g. see Reference 15). If the oscillatory signature lasts for only a relatively short period each time it occurs, it may be easily masked if spectral analysis is applied over a long observation interval. Also, capturing such a signature may require a high sampling rate that may not be acceptable to sample the signal for a long period of time; a flexible sampling scheme is desirable. For the reasons to be shown next, wavelet transform is more suitable in such a case. It can employ a long window and low sampling rate for a low-frequency component in the signal, and at the same time a short window and high sampling rate for a high-frequency component. Furthermore, wavelet transform exhibits time–frequency localisation, which gives a more precise description of the signal, providing more informative data for condition monitoring.

It is true that condition monitoring is usually applied to steady-state signals, even if the signature carried by the signals is of an intermittent nature. However, it has been realised that sometimes the target signature can be too difficult to separate from the dominant fundamental signal. For example, a broken rotor bar fault in an induction machine causes side band current components whose frequencies are only slightly different from the fundamental frequency, the difference depending on the slip, which is normally very small in the steady state. It has recently been suggested that a more effective method is to detect the current signatures during start-up of the machine when slip is large [16,17]. However, the transient current during start up is not a stationary and periodic signal. Conventional spectral analysis would give erroneous results in this case. Again, wavelet transform, due to its time–frequency localisation property, can be a more suitable technique for extracting the signatures from the transient current.

Fourier analysis uses sine waves of different amplitudes, frequencies and phase angles to synthesise a periodic signal. Wavelet transform expands this concept by considering that a time domain signal, not necessarily periodic, can be reconstructed using a series of small waveforms that can be transitioned in time and scaled in amplitude. Such small waveforms, which are called wavelets, no longer need to be periodic, like sine waves. As an example, for a time signal $f(t)$, and a chosen basic mother wavelet function $\psi(t)$, the corresponding wavelet transform, $W(a,b)$, is defined as:

$$W(a,b) = \frac{1}{\sqrt{a}} \int_{-\infty}^{\infty} f(t)\psi^* \left(\frac{t-b}{a}\right) dt, a > 0, -\infty < b < \infty \qquad (5.16)$$

where a is the scaling factor of time, b is the time-shifting parameter in the mother wavelet function and ψ is defined below. The mother wavelet function is usually chosen as an oscillatory waveform that decays in both directions from the centre of the wavelet, as shown in Figure 5.9. So parameters a and b in (5.16) therefore scale the

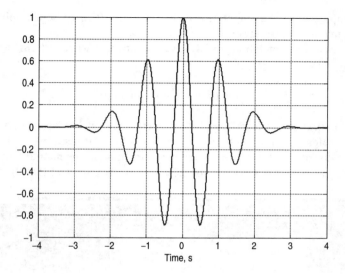

Figure 5.9   A mother wavelet function

mother wavelet frequency and shift the centre of the mother wavelet respectively. For a given instant $t = b$, if the signal $f(t)$ contains significant oscillation at a frequency coinciding with $1/a$, then the integration shown in (5.16) will result in a significant output. Because the mother wavelet decays in both directions away from the centre $t = b$, what happens to the signal $f(t)$ a long time before $t = b$ or after $t = b$ does not matter. Therefore wavelet transform provides the local frequency information in signal $f(t)$ around $t = b$. As parameter b is varied, the $f(t)$ is scanned through in terms of time. As parameter a is changed, signatures at different frequencies in the signal are revealed.

The decaying rate of the mother wavelet is usually selected to depend on the scaling factor a. If a is small, so that the corresponding frequency is high, then the decaying rate can be fast. If a is large, corresponding to a low frequency, then the decaying rate is slow. In this way, wavelet transform exhibits the property of multiple resolutions. The low-frequency information, which is present for a relatively long period of time in the signal, and the high-frequency information, which exists for a short period of time, can be obtained simultaneously. The most commonly used mother wavelet function, whose waveform is partially (real part) shown in Figure 5.9, is the Gaussian function:

$$\psi(t) = e^{-t^2/2}[\cos(2\pi t) + j\sin(2\pi t)] \tag{5.17}$$

It is obvious that wavelet transform as defined in (5.16) can be calculated numerically in a way similar to that shown in Figure 5.7, further taking into account the complex number nature of the mother wavelet function. It is, however, important to note that wavelet transform may involve different sampling rates and different observation lengths. It is therefore important to keep synchronisation in the sampling scheme. As

$\psi(t)$ is complex, from the output of wavelet transform $W(a, b)$ both amplitude and phase information can be retrieved.

For the sake of completion, (5.18) gives the inverse wavelet transform where $C(\psi, v) = \int_0^\infty [\psi^*(v)\psi(v)/v]dv$:

$$f(t) = \frac{1}{C} \int_0^\infty \int_{-\infty}^\infty \frac{1}{a^2\sqrt{a}} W(a, b)\psi(\frac{t - b}{a})db \cdot da \tag{5.18}$$

Figure 5.10(a) shows a measured starting current signal of an induction motor taken from References [18] and [19]. It was found that transient torsional vibration was excited during the direct on-line starting of the motor. The natural frequency of the mechanical shaft was known to be 22 Hz. Research showed that the torsional

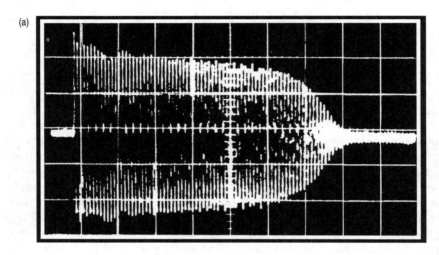

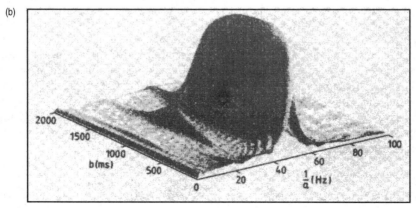

Figure 5.10   Monitoring the onset of torsional vibration during the start up of a downhole motor pump set [Taken from Yacamini, Smith and Ran [18,19]]

vibration could be indicated in the motor stator current as side band components at 28 Hz and 72 Hz assuming a supply frequency of 50 Hz. As the starting current was transient and non-stationary, wavelet transform was applied to the current signal and the result is shown in Figure 5.10(b). The development of the 28 Hz component is clearly observed while the 72 Hz component is shadowed by the fundamental [18].

5.7 Conclusion

This chapter has described the signal processing methods now available for computer monitoring electrical machines. The following five chapters will describe the major temperature, chemical, mechanical and electrical techniques available for monitoring.

5.8 References

1. Madisetti V.K. and Williams D.B. (eds.) *The Digital Signal Processing Handbook*. Boca Raton: CRC Press; 1998. pp. 1-21–1-23.
2. Durocher D.B. and Feldmeier G.R. (2004). Predictive versus preventive maintenance. *IEEE Industry Applications Magazine* 2004; (September/October): 12–21.
3. Flankin G.F., Powell J.D. and Workman M. *Digital Control of Dynamic Systems*. Menlo Park: Addison-Wesley Longman; 1998. pp. 160–66.
4. Mohan N., Undeland T.M. and Robbins W.P. *Power Electronics – Converters, Applications and Design*. New York: John Wiley & Sons; 2003. pp. 225–30.
5. Jintakosonwit P., Fijita H. and Akagi H. Control and performance of a fully-digital-controlled shunt active filter for installation on a power distribution system. *IEEE Transactions on Power Electronics* 2002; **PE-17**: 132–40.
6. Arthur N. and Penman J. Induction machine condition monitoring with higher order spectra. *IEEE Transactions on Industrial Electronics* 2000; **IE-47**: 1031–41.
7. Onadera S. and Yamasawa K. Electromagnetic vibration analysis of a squirrel-cage induction motor. *IEEE Transactions on Magnetism* 1993; **M-29**; 2410–12.
8. Yazici B. Statistical pattern analysis of partial discharge measurements for quality assessment of insulation systems in high-voltage electrical machinery. *IEEE Transactions on Industry Applications* 2004; **IA-40**: 1579–94.
9. Kang P. and Birtwhistle D. Condition monitoring of power transformer on-load tap-changers, part 1: automatic condition diagnostics. *IEE Proceedings Generation Transmission Distribution* 2001; **148**, 301–306.
10. Wismer N.J. *Gearbox Analysis using Cepstrum Analysis and Comb Liftering*. Bruel & Kjaer Application Note. Naerum: Bruel & Kjaer; 2002.
11. Morris, A.S. *Measurement & Instrumentation Principles*. Oxford: Butterworth-Heinemann; 2001.
12. Stack J.R., Habetler T.G. and Harley R.G. Bearing fault detection via auto-regressive stator current modeling. *IEEE Transactions on Industry Applications* 2004; **IA-40**: 740–47.

13. Tavner P.J., Gaydon B.G. and Ward D.M. Monitoring generators and large motors. *IEE Proceedings, Part B, Electric Power Applications* 1986; **133**: 169–80.
14. Flordin P. *Time-Frequency Time Scale Analysis*. New York: Academic Press; 1999.
15. Lin J. and Zuo M.J. Gearbox fault diagnosis using adaptive wavelet filter. *Mechanical Systems and Signal Processing* 2003; **17**: 1259–69.
16. Zhang Z., Ren Z. and Huang W. A novel detection method of motor broken rotor bars based on wavelet ridge. *IEEE Transactions on Energy Conversion* 2003; **EC-18**: 417–23.
17. Douglas H., Pillay P. and Ziarani A.K. A new algorithm for transient motor current signature analysis using wavelets. *IEEE Transactions on Industry Applications* 2004; **IA-40**: 1361–8.
18. Yacamini R., Smith K.S. and Ran L. Monitoring torsional vibrations of electromechanical systems using stator currents. *Journal of Vibration and Acoustics* 1998; **120**; 72–9.
19. Ran, L., Yacamini R. and Smith K.S. Torsional vibrations in electrical induction motor drives during start up. *Journal of Vibration and Acoustics* 1996; **118**: 242–51.

Chapter 6
Temperature monitoring

6.1 Introduction

As has been described in Chapter 2, the limits to rating of electrical machines are generally set by the maximum permissible temperature that the insulation can withstand. Indeed the performance testing of machines, before they leave a manufacturer's works, is dominated by the measurement of winding or embedded temperatures and the need to achieve temperature rises within the appropriate standards. The measurement of temperature therefore has an important place in the monitoring of electrical machines and the following describes how this can be done using the measurement techniques described in Chapter 4.

There are three basic approaches to temperature monitoring.

1. To measure local temperatures at points in the machine using embedded temperature detectors.
2. To use a thermal image, fed with suitable variables, to monitor the temperature of what is perceived to be the hottest spot in the machine.
3. To measure distributed temperatures in the machine or the bulk temperatures of coolant fluids.

These approaches demonstrate the fundamental difficulty of thermal monitoring, which is resolving the conflict between point temperature measurements that are easy to make, but give only local information, and bulk temperature measurements that are more difficult and run the risk of overlooking local hot-spots.

The following three sections show how these approaches can be applied practically.

6.2 Local temperature measurement

This can be done using thermocouples, resistance temperature detectors or embedded temperature detectors, whose characteristics are described in Chapter 4. To monitor the active part of the machine they are usually embedded in the stator winding and in the stator core. They can also be located in the bearings to detect hot running. The choice of location requires careful consideration during the specification stage of the machine. For example temperature detectors embedded in the stator winding need to be located close to its hottest part, which may be in the slot portion or end-winding portion, depending on the thermal design of the machine. Alternatively, for a machine with an asymmetrical cooling arrangement they should be located at the hottest end of the machine. Some guidance as to where embedded temperature detectors should be located is given by Chalmers [1] and this is summarised in Figure 6.1. It should be

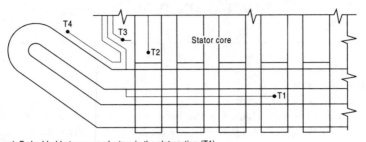

a) Embedded between conductors in the slot portion (T1)
   Embedded in core pack laminations (T2)
   Mounted on potentially hot components such as pressure plates (T3)
   Embedded on conductors in the end winding portion (T4)

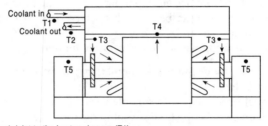

b) Water or air inlet to the heat exchanger (T1)
   Water or air outlet from the heat exchanger (T2)
   Re-entry air to the machine from the heat exchanger (T3)
   Exhaust air from the machine to the heat exhanger (T4)
   Bearing temperatures (T5)

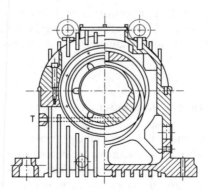

c) Bearing temperature (T)

Figure 6.1   Location of temperature detectors in electrical machines

noted that when embedded temperature detectors are fitted in bearings, precautions must be taken to ensure that the bearing insulation is not breached.

The weakness of these methods is that thermocouples and resistance temperature detectors are metallic devices and cannot be located on the hottest active component, the winding copper, because they need electrical isolation. On a winding the devices have to be embedded in the insulation at some distance from the copper itself (see Figure 6.2(a)). As a result, the measured temperature will not necessarily be that of the winding itself but an image of it, as shown in Figure 6.2(b). The heat flow per unit area, Q, through the insulation system can be described by simple conduction equations as follows:

$$Q = h(T_s - T_g) = \frac{k}{t_2}(T_s - T_t) = \frac{k}{t_1}(T_c - T_t) \tag{6.1}$$

where t is insulation thickness, T is temperature, k is the heat transfer coefficient through an insulating material, and h is the heat transfer coefficient from the insulation surface.

Eliminating T_s between the first two expressions gives

$$T_t = T_g + Q\left(\frac{t_2}{k} + \frac{1}{h}\right), \tag{6.2}$$

therefore

$$T_t = T_g + (T_e - T_t)\left\{\frac{(t_2 + (k/h))}{t_1}\right\} \tag{6.3}$$

and

$$T_1 = \frac{T_g + T_c\left\{\frac{(t_2 + (k/h))}{t_1}\right\}}{1 + \left\{\frac{(t_2 + (k/h))}{t_1}\right\}} \tag{6.4}$$

so

$$T_1 \approx T_c \text{ if } T_g \ll T_c \text{ and } \frac{(t_2 + (k/h))}{t_1} \gg 1; \text{ that is, if } t_2 + \frac{k}{h} \gg t_1 \tag{6.5}$$

So the measured temperature T_t will approach the temperature of the hottest active component T_c if the thickness of insulation, t_2, applied over the ETD is sufficient compared to the main insulation. This problem does not occur for devices embedded in the slot portion between two conductors, as shown in Figure 6.2(c), where there is a low heat flux between the active copper parts. But it is an important difficulty when monitoring end-winding temperatures, such that the thickness of overtaped insulation, t_2, needs to be substantial if sensible readings are to be obtained.

It would be very desirable to develop a temperature-monitoring device that can be affixed to a high-voltage winding and give electrical isolation. Such a device was developed in the 1980s for power transformers, comprising a small phial of liquid that had a high vapour pressure that varied widely with temperature. The phial

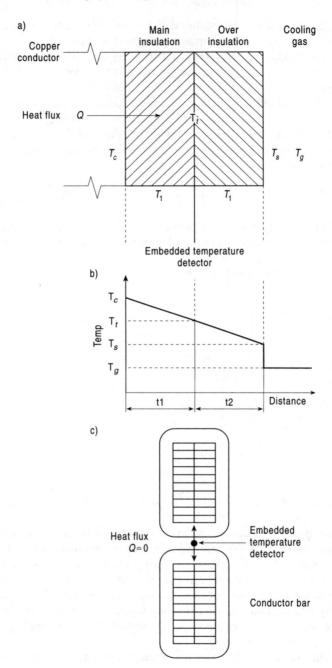

*Figure 6.2   Effect of embedding a temperature detector away from an active part.
(a) Asymmetrically embedded temperature dectector. (b) Temperature
dectection from (a). (c) Symmetrically embedded temperature detector.*

was affixed to the high voltage winding whose temperature was required. The pressure in the phial was then monitored through a non-conducting silicone rubber tube and the temperature was derived electronically from the pressure measurement. The device was applied, on a trial basis, to operational transformers where it performed satisfactorily.

More widespread modern methods of measuring temperature in high-voltage components have also been developed using fibre-optic techniques, as described in Chapter 4 and shown in Figure 6.3. A particular design using the dependence of the polarisation of light on the temperature of a material is described by Rogers [2,3].

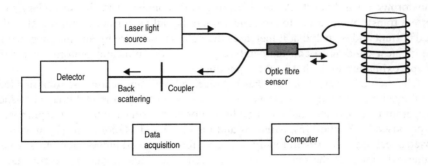

Figure 6.3   Principle of fibre-optic temperature sensing

Light from a laser is transmitted to the device and is passed through a polarising prism before being launched into a fibre maintained at the temperature to be measured. This material introduces a rotation of the light beam that is dependent on temperature. The beam is reflected back through the fibre and polariser and is relaunched with a light amplitude proportional to the polarisation that has taken place in the fibre. By arranging two passes through the fibre the polarisation is rendered insensitive to electrical or magnetic field effects.

The temperature measurements described so far have all been on stationary parts of the machine and on many machines the thermal design is stator-critical, so the hottest spot will be located there. But many machines are rotor-critical, particularly larger induction motors, because under stall conditions the rotor losses are very large, and the temperatures can rise rapidly to values that could damage the rotor integrity. On such machines there may well be no apparent deterioration after one or two stalls, but should this occur repeatedly there will be a weakening of rotor bars and/or end rings that may result in premature mechanical failure. In the past there have been various crude methods of measuring rotor temperatures for experimental purposes, using heat-sensitive papers or paints, or thermocouples connected through slip rings. Until recently, however, there has not been a method sufficiently reliable to use for monitoring purposes. Siyambalapitiya *et al.* [4] have described a device for monitoring eight thermocouples, multiplexing the signal on the rotor, then optically coupling to the stator and decoding in a digital signal processor.

6.3   Hot-spot measurement and thermal images

Local temperature measurements give the machine operators considerable confidence that they knows the operating temperature of key points in the machine, but there is always the nagging suspicion that temperature detectors may not be located at the hottest point. This problem has long been recognised in power transformers where it is extremely difficult to obtain even embedded winding temperatures, because of the need for extra-high voltage isolation and the great thickness of electrical insulation necessary, so thermal images of the hot spot temperature are used. The thermal image consists of a dial-type thermometer with its bulb immersed in the region where the transformer oil is hottest. A small heating coil, connected to the secondary of a current transformer, serves to circulate around the bulb a current proportional to the load current and is such that it increases the bulb temperature by an amount equal to the greatest winding-to-coil temperature gradient. The indicator therefore registers an approximation to the hot-spot temperature.

The thermal image technique has not received wide application on rotating electrical machines, although it deserves to. The availability of a thermal image hot-spot temperature of machine could be used for motor monitoring and protection purposes, as proposed by Zocholl [5], Milanfar and Lang. [6] and Mellor *et al.* [7], who proposed a technique for small, totally enclosed, forced-cooled induction motors where a thermal model of the machine is configured in a digital signal processor that is fed with signals proportional to the ambient air temperature and the stator winding current. The model can calculate the predicted temperatures at a variety of key points in the machine. Stator core, stator winding slot, or stator end-winding representation must be programmed solely using the design information for the machine. The instrument is designed to produce an analogue voltage that is proportional to several temperatures at the hottest points.

Mellor *et al.* have used the device on two totally enclosed fan-cooled machines and compared the predictions of the thermal image with the actual measured temperatures in the stator end winding that was found to be the hottest point in this design of the machine. Figure 6.4 shows the comparison of results for these two machines after they had been put through very severe duty cycles and it can be seen that the results are extremely good. The device has been designed to be part of the thermal protection of a motor but could equally well be used for monitoring a machine for operational purposes, particularly, for example, on a crucial machine located in an inaccessible position where hot-spot measurements may be difficult to obtain.

6.4   Bulk measurement

In the electrically active part of the machine, even when hot-spot locations are known or hot-spot temperatures can be surmised from a thermal image, there is still a desire to obtain a bulk indication of the thermal state of the machines. This can be found from the measurement of the internal and external coolant temperature rises, obtained from thermocouples located, for example, as shown in Figure 6.1. This is done on

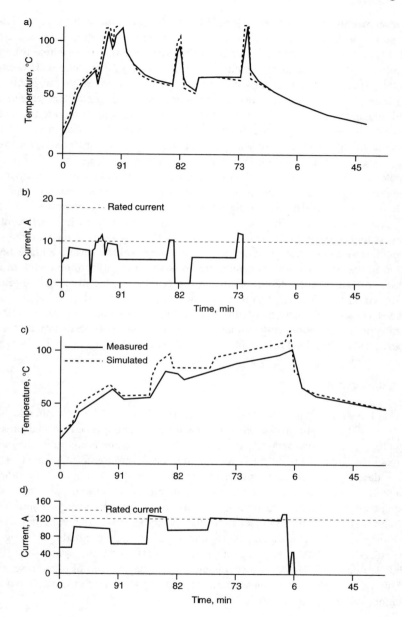

*Figure 6.4   Comparison between measurements and the predictions of a ther-
mal image of an electrical machine. [Taken from Mellor et al. [7]]
(a) Comparison for 5.5 kw induction motor. (b) Duty cycle for (a).
(c) Comparison for 7.5 kw induction motor; Duty cycle for (c).*

most larger machines but it is normal for coolant temperatures to be displayed and rare for the point values to be subtracted to give the temperature rise directly. An increase in temperature rise from such a device would clearly show when a machine is being overloaded or if the coolant circuits are not performing as they should. However, the method is insensitive to localised overheating in the electrically active parts of the machines, therefore considerable effort has been devoted, as an alternative to the thermal image, to devising methods whereby single indications of high temperature can be obtained from a device that is embedded in the bulk of the machine.

Lengths of signal cable using heat-sensitive semiconducting material as the insulation have been proposed but most effort has been devoted to the use of optical fibres. Brandt and Gottlieb [8] have described various methods, including simple point measurements on high-voltage components, using the optical fibre for isolation purposes. They have also described how, using the temperature sensitive properties of fibre optics, a continuously sensitive fibre could be embedded in the machine, adjacent to the high-voltage copper, to detect localised overheating anywhere in the winding and yet provide a single indication. The method proposed would utilise the black body radiation in the optical fibre, alongside the hot-spot, being transmitted back to the detector and being used to determine the hottest point along the fibre's length. Such a device would therefore need to be embedded in the machine during manufacture and as yet a practicable instrumentation scheme has not been devised. An interesting modern approach to monitoring temperature in a variable speed induction motor by Chen *et al.* [9] is described briefly in a later chapter.

6.5   Conclusion

This chapter shows that temperature measurement can yield very valuable bulk indications of the condition of an electrical machine using simple sensors and narrow-bandwidth (<1 Hz), low-data-rate signals and, because temperature limits the rating of a machine, over-temperature is a valuable condition-monitoring signal. Temperature detection has repeatedly been shown to be an effective global monitoring technique for electrical machines, but has been neglected as a monitoring method.

Temperature measurement is usually done in traditional and rather antiquated ways, and there are some simple changes that could be made in existing practice to make more sense of it. These changes are generally in the area of signal processing and in particular the importance of presenting temperature rises to the operator, rather than absolute temperature. There are also advances in the application of modern sensors, which will allow temperature measurements to be made closer to the active parts of a machine, and these should be exploited.

6.6   References

1. Chalmers B.J. *Electrical Motor Handbook*. 1st ed. London: Butterworths; 1988.
2. Rogers A.J. Optical temperature sensor for high voltage applications. *Applied Optics* 1982; **21**: 882–85.

3. Rogers, A.J. Distributed optical-fibre sensing. *Measurement Science & Technology* 1999; **10**: R75–R99.
4. Siyambalapitiya D.J.T., McLaren P.G. and Tavner P.J. Transient thermal characteristics of induction machine rotor cage. *IEEE Transactions on Energy Conversion* 1988; **EC-3**: 849–54.
5. Zocholl S.E. Motor analysis and thermal protection. *IEEE Transactions on Power Delivery* 1990; **PD-5**: 1275–80.
6. Milanfar P. and Lang J.H. Monitoring the thermal condition of permanent-magnet synchronous motors. *IEEE Transactions on Aerospace and Electronic Systems* 1996; **AES-32**: 1421–29.
7. Mellor P.H., Roberts D. and Turner D.R. Lumped parameter thermal model for electrical machines of TEFC design. *IEE Proceedings Part B, Electric Power Applications*, 1991; **138**: 205–218.
8. Brandt G.B. and Gottlieb M. Fiber-optic temperature sensors using optical fibers. *Abstracts of Papers – American Chemical Society* 1982; **184**: 64-ANYL.
9. Chen S.E., Zhong E. and Lipo T.A. A new approach to motor condition monitoring in induction motor drives. *IEE Transactions on Industry Applications*. **1A-30(4)**: 905–911.

Chapter 7
Chemical monitoring

7.1 Introduction

The insulating materials used in electrical machines are complex organic materials, which, when they are degraded by heat or electrical action, produce a very large number of chemical products in the gas, liquid and solid states. Lubrication oils also carry not only the products of their own degradation but also those from the wear of the bearings and seals they cool and lubricate. Techniques to provide early warning of the deterioration of electrical machines should include measurement of these complex degradation products where they can be accessed in the machine.

7.2 Insulation degradation

What are the mechanisms by which insulation can be degraded? Well, the table in the Appendix shows how important excessive temperature and therefore thermal degradation is as a failure mode and this is largely determined by the thermal performance of the insulation. The insulation mainly consists of organic polymers, either natural forms such as bitumen, or more commonly nowadays, synthetic epoxy resins. Chapter 2, Section 2.7.2 has described the majority of the failure modes associated with insulation degradation. The thermal degradation of these materials is a complex process. However, as the temperature of the insulation rises above its maximum permitted operating value, $c.$ 160 °C, volatiles used as solvents in the insulation manufacture start to be driven off as gases. Then the heavier compounds making up the resin may reach their boiling point. The gases so produced are generally the heavier hydrocarbons such as ethylene. As the temperature rises further, above 180 °C, chemical decomposition of the resin components starts. A supersaturated vapour of the heavier hydrocarbon decomposition products then forms in the cooling gas close to the high temperature area of insulation. Rapid condensation of that vapour occurs as the cooling gas leaves the hot area producing condensation nuclei that continue to grow in size by further condensation until they reach a stable droplet size. These droplets, usually called particles, are of submicron size and form what would commonly be called smoke. The precise materials given off depend primarily on the insulation material being heated but also on the cooling gas of the machine. The binder material of the insulation, whether it be wood, paper, mica or glass fibre, can usually withstand much higher temperatures, but eventually as 400 °C is reached, they start to degrade and char, releasing gases such as carbon monoxide and carbon dioxide, drawing oxygen from the cooling gas, if it is air, or from the degradation of the complex hydrocarbon in the resin. Figure 7.1 shows a piece of phenol-impregnated wood insulation from a turbine generator, progressively raised to higher temperatures in hydrogen. Up to 300 °C one can see the effects of the resin material being decomposed and driven off, but the wood binder still retained its strength. Above 400 °C the binder has degraded by charring and no longer has significant mechanical strength. Pyrolysing activity therefore gives rise to a wide range of liquid droplets, solid particulates and gases, which together make up the smoke being driven off from the insulation.

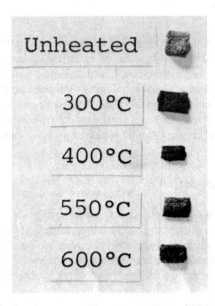

Figure 7.1   Heating of phenolic-impregnated wood insulating material

Electrical discharge activity, within or adjacent to the insulation system, can also degrade the insulation releasing particulate and gaseous chemical degradation products. The very high temperature associated with sparking breaks down the hydrocarbon compounds in the insulation to form acetylene. It also breaks down the oxygen in the cooling gas, if it is air, to give ozone. Furthermore, continuous discharge activity gradually carbonises and erodes the insulating material to produce, on a smaller scale, the degradation products that result from more widespread overheating.

7.3   Factors that affect detection

Before considering the different methods of detecting chemicals within an electrical machine it is necessary to understand the factors that affect the detectability. Consider

Figure 7.2(a), which shows the machine within its enclosure, the rate of change of concentration of a detectable substance can be determined by the following equation:

Rate of change of concentration of a detectable substance in the cooling circuit of a machine = (specific production rate of detectable substance - leakage rate from machine)/machine volume

$$\frac{dC}{dt} = \left\{ \frac{\dot{v} - \frac{VC}{\tau_r}}{V} \right\} \tag{7.1}$$

where V is the machine volume, C is the volumetric concentration of the substance concerned, τ_r is a leakage factor and $\dot{v}$ is the volumetric rate of production of the detectable substance.

When leakage is low, τ_r is effectively a residence time constant for the substance in the machine enclosure. The rate of production, v, is related to the volume of material being overheated and its chemical composition. This variable is in fact a function of time $v(t)$. So the equation could be rewritten:

$$\frac{dC}{dt} + \frac{C}{\tau_r} = \frac{\dot{v}(t)}{V} \tag{7.2}$$

$$\left(D + \frac{1}{\tau_r} \right) C = \frac{\dot{v}(t)}{V} \tag{7.3}$$

The complementary function for this equation is

$$C = A \exp\left(-\frac{t}{\tau_r} \right) \tag{7.4}$$

The particular integral is

$$C = \frac{(\dot{v} - \dot{v}_b)\tau_r}{V} \tag{7.5}$$

where $\dot{v}(t) = \dot{v} + \dot{v}_b$, $\dot{v}$ is a step increase in the volumetric rate of production and $\dot{v}_b$ is the background rate of production of the substance.

When $t = 0, C = C_b$ (the background concentration), then

$$C_b = \frac{\dot{v}_b \tau_r}{V} \text{ and } A = \frac{\dot{v} \tau_r}{V} \tag{7.6}$$

$$\therefore C = \frac{\dot{v} \tau_r}{V} \left(1 - \exp\left(\frac{t}{\tau_r} \right) \right) + \frac{\dot{v}_b \tau_r}{V} \tag{7.7}$$

By studying (7.7) it is possible to determine how the detectable concentration depends upon the design of the machine, the nature of the overheating it is suffering from and the concentration of material involved. Figure 7.2(b) will help to explain this. A machine with a tightly sealed and pressurised cooling system, such as a turbine generator, will have a long residence time (τ_r) of many hours, the background concentration level of the detectable substance (C_b) will be high. On the other hand the concentration level, which can be reached after an extended period of overheating, will also be correspondingly high. But if the length of the overheating incident (τ_o) is short compared to the residence time, say a few minutes, then the concentration will not build up to a significant level compared to the background. A machine with an open cooling circuit will have a short residence time (τ_r) of perhaps merely a few seconds. So the background concentration will be low and there will need to be a large volumetric production (V) from an overheating incident to produce a large increase in concentration (C), but the concentration level will respond rapidly to any overheating.

Detectability of overheating depends upon:

1. a large signal-to-noise ratio; that is, the magnitude of the indication, (X in Figure 7.2(b)), must be large compared to the background, C_b;
2. a long duration of indication.

The larger the signal-to-noise ratio of the indication and the longer its duration the easier it will be to detect. These two conditions can be considered mathematically:

For a large signal-to-noise ratio, (1) above, $\dot{v} \gg \dot{v}_b$ and $X \gg (\dot{v}_b \tau_r)/V$. Now

$$X = \frac{\dot{v}}{V} \tau_r \left(1 - \exp \left(\frac{\tau_0}{\tau_r} \right) \right) \tag{7.8}$$

where τ_0 is the duration of the overheating incident. So

$$\left(1 - \exp \left(-\frac{\tau_0}{\tau_r} \right) \right) \gg \frac{\dot{v}_b}{\dot{v}} \tag{7.9}$$

$$\left(1 - \frac{\dot{v}_b}{\dot{v}} \right) \gg \exp \left(-\frac{\tau_0}{\tau_r} \right) \tag{7.10}$$

That is $\dot{v}$ must be $\gg \dot{v}_b$ and $\tau_0 > \tau_r$, or $\dot{v}$ must be $> \dot{v}_b$ and $\tau_0 \gg \tau_r$. For a long duration of indication, (2) above, $\tau_0 \gg \tau_r$.

The time constant, τ_0, depends on

- the type of fault causing the overheating,
- the extent of the fault,
- the nature of the material being overheated,
- the nature of the substance being released and detected.

So, for example, a small intermittent shorted turn in the rotor winding of a large turbine generator will produce heating for only a few minutes and the increase in concentration of detectable substances in that time will be small because of the long residence time, τ_r, in such a machine. On the other hand overheating in a terminal connection or due to excessive stray losses will operate for many hours and produce a substantial indication.

The effect of the substance being detected can be considered as follows. An insulation material that is heated at a steady but relatively low temperature of say 190 °C will produce a considerable amount of hydrocarbon gases over a long period of time, up to 2–3 hours, until those gases are all driven off. Particulates will also be formed but these will have a short lifetime in the enclosure because of recombination

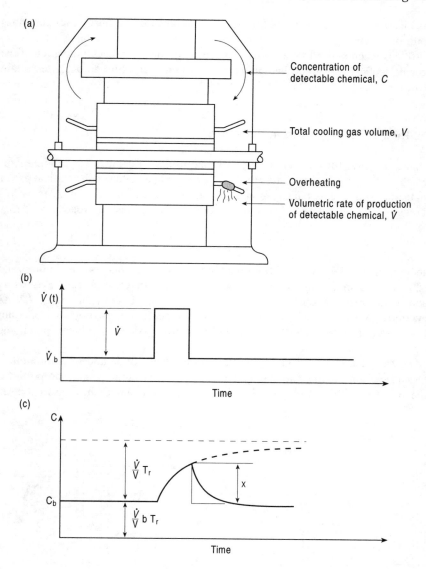

Figure 7.2 *The concentration of detectable chemicals in coolant gas of an electrical machine. (a) Cross-section of machine. (b) Step change in the rate of production of chemical. (c) Response of the concentration of detectable chemical to a step change in the rate of production of that chemical.*

and condensation, say 10–15 minutes. If the insulation is raised to higher temperatures the production of copious quantities of gases will take place over a much shorter period of time and will stop before charring commences. If the machine is air-cooled, large amounts of carbon monoxide and carbon dioxide will also be produced, if overheating takes place over a long period of time, because production of those gases continues even during charring.

All these factors need to be taken into consideration when deciding which of the following techniques should be applied to a particular machine to detect a certain type of fault.

7.4   Insulation degradation detection

It can be concluded from Sections 7.1–7.3 that insulation degradation can be monitored chemically by detecting the presence of particulate matter in the coolant gas or by detecting simple gases like carbon monoxide and ozone, or more complex hydrocarbon gases like ethylene and acetylene. Let us consider these approaches in turn.

7.4.1   Particulate detection: core monitors

Detecting the smoke given off from degrading insulation appears the simplest and most general of all techniques, since proprietary smoke detectors already exist using an ion chamber to detect the smoke particles. An example is shown in Figure 7.3. As the cooling gas of the machine enters the ion chamber it is ionised by a weak radioactive source. The gas then flows through an electrode system to which a polarising voltage is applied.

The free charges in the gas are collected on the electrode and flow through an external electrometer amplifier circuit, which produces an output voltage proportional to the ion current. When heavy smoke particles enter the chamber they too are ionised

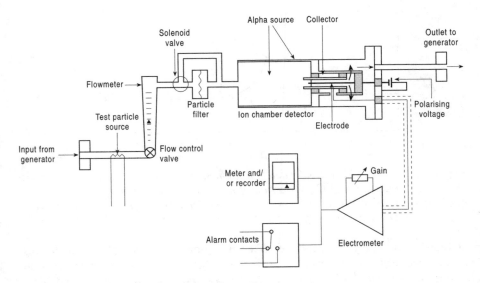

Figure 7.3   Diagram of a basic core monitor. [Taken from Skala [1]]

and their greater mass implies a lower mobility compared to the gas molecules, so as they enter the electrode system the ion current reduces. The smoke is therefore detected by a reduction in the output voltage from the electrometer amplifier. Skala [1] described an ion chamber specifically designed to detect the products of heated insulation and this was applied to a large turbine generator by Carson *et al.* [2].

The primary impetus for this work was the need to provide early warning of core faults referred to by Tavner *et al.* [3,4], which the larger sizes of turbine generators started to experience in the early 1970s. A core fault can involve substantial quantities of molten stator steel and, hitherto, the fault could only be detected when the melt burnt through the stator winding insulation and caused an earth fault. It was hoped that the core monitor could detect the degradation of the insulation between the steel laminations at an earlier stage in the fault. The lifetime of pyrolised particles in the closed hydrogen cooling circuit of a large generator is 15–30 min after which time the particulates are deposited onto the exposed surfaces of the machine. So a single instance of insulation overheating should lead to a reduction of core monitor ion current for a period of time of this order. Figure 7.4 shows typical core monitor responses. When a core fault is occurring the overheating continues over a longer period and it has been shown by Carson *et al.* [2] that the core monitor does respond to this and other forms of insulation overheating. Core monitors are available from several manufacturers. The sensitivity of the device depends upon the ion chamber design but experimental figures for the monitor described in Reference 2 show that it will produce a response ranging from 85–95 per cent of full-scale deflection when 100 cm^2 of lamination insulation is pyrolised, depending on the material. An area is quoted because the production of particulates is primarily a surface effect.

The device, however, does have some practical difficulties.

1. The monitor output fluctuates with cooling gas pressure and temperature.
2. The monitor responds to oil mist that may be present in the circuit of any hydrogen-cooled machine due to faulty hydrogen seals [5].
3. The monitor is non-specific; that is, it cannot distinguish between the materials being overheated.

Items (1) and (2) affect the background signal from the monitor, which any signal due to damaging overheating must exceed. Figure 7.4(c) shows a typical core monitor trace from a machine affected by oil mist. Item (3) affects the attitude of a machine operator to an alarm from the core monitor, since there will be less confidence in the monitor if it is not known from what part of the machine the detection originated.

A more advanced monitor, described by Ryder *et al.* [6], has been devised to overcome problems (1) and (2) by using a differential technique, Fig. 7.5. Ryder's monitor consists of two identical ion chambers in series in the gas flow line with an intermediate particulate filter between them. The monitor displays the difference between the ion currents in the two chambers and thereby eliminates fluctuations due to pressure and temperature.

The sensitivity of a core monitor to oil mist can be reduced if the ion chamber is kept at an elevated temperature. It has been proposed that oil mist is only produced by overheating, so that its detection may be useful. The use of heated ion chambers

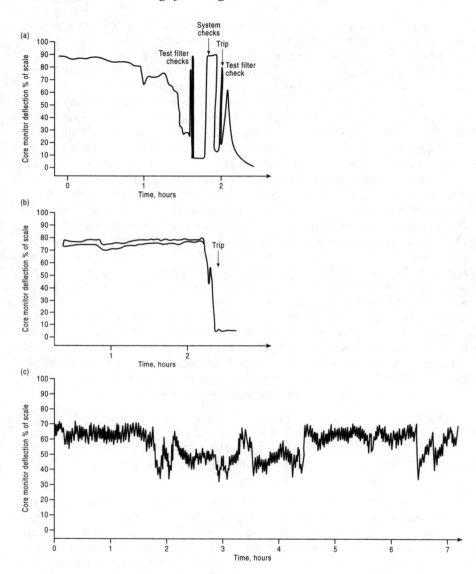

Figure 7.4   Typical core monitor responses. (a) Machine with overheating conductor bar. (b) Machine with a core fault. (c) Machine with no overheating but heavy oil contamination.

was not initially encouraged, however, current thinking is that heated ion chambers are essential for reliable detection. However, the amount of oil in a turbine generator casing varies widely and can be particularly high. In this case it has been found that there can be frequent false core monitor alarms, so the use of a heated ion chamber gives a significant advantage. In order to completely vaporise an oil mist it is

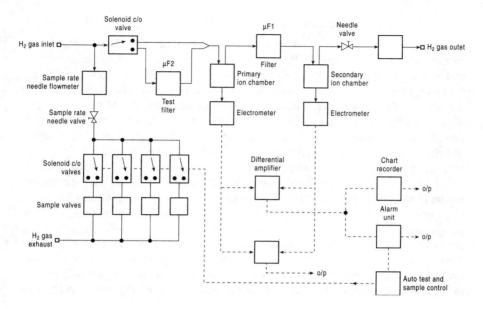

Figure 7.5   *Diagram of a differential core monitor with heated ion chambers, see*
              Tavner et al. [4]

necessary to raise the ion chamber temperature above 120 °C. The monitor described
in Reference 6 has heated ion chambers and the authors' experience, using these set
to 120 °C, was that they gave adequate protection against spurious oil mist indication. Using a heated ion chamber also means that some of the droplets produced by
overheating will be vaporised, or at least reduced in size, and this must result in a
consequential loss of sensitivity. However, laboratory tests can quantify this loss of
sensitivity, which at 120 °C has been shown to be 20 per cent, Braun and Brown
have shown more recently by careful tests, the deleterious effects of using heated ion
chambers on the core monitor sensitivity [7].

Use of core monitors in the early 1970s was advocated in the US as a panacea for
the early detection of major core and winding faults. Since then, however, although
there have been some notable detection successes, such as those traces shown in
Figures 7.4(a) and (b), there have also been false alarms, many caused by oil mist
(see Figure 7.4(c)).

The authors are not aware of the core monitor being used on air-cooled machines,
or machines without a closed cooling circuit at all, although apart from the short time
constant of the indication from the monitor there seems to be no reason why it should
not be used for these applications. Experience has shown that the core monitor cannot
be relied upon, on its own, to give incontrovertible evidence of an incipient fault. It
is a valuable device that does detect pyrolised insulation but its indications need to
be considered alongside those of other monitoring devices. In particular the core
monitor needs to be complemented by an off-line technique to chemically analyse
the particulate material causing the detection, as described in the following section.

7.4.2   *Particulate detection: chemical analysis*

Many authors have advocated taking a sample of the particulate material when a core monitor indicates an alarm [2]. In order to collect a detectable amount of particulate matter within a short time, it is necessary to have a very large gas flow-rate through the filter. This is achieved by venting the pressurised casing of the machine through the filter to the atmosphere. There is not such close agreement about the method of analysis however. Carson *et al.* [2] described a method whereby the pyrolysis products are collected upon a small charge of silica gel and are then released into a gas chromatograph upon the application of strong heat. This technique is applicable only when sampling is carried out immediately upon detection of local overheating. This is because it is specially designed to collect the particulates and heaviest pyrolysis products, which are present for only a limited time in the gas, sometimes for only a few minutes. There is also a problem with the gas chromatographic analysis of the pyrolysis products as these products contain a very large number of different organic compounds and the resultant chromatogram is difficult to interpret. Figure 7.6 taken from Reference 8 gives an example. The most arduous test would be to distinguish the products of pyrolised insulation from the overheated oil that is generally present in any electrical machine. Further work on this from Dear *et al.* [8] on the gas released

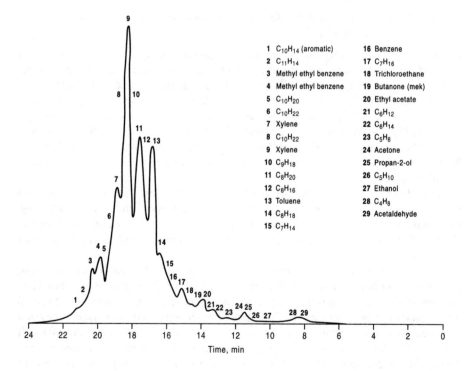

Figure 7.6   Gas chromatogram of generator hydrogen impurities described from a Tenax GC pre-column. [Taken from Dear et al. [8]]

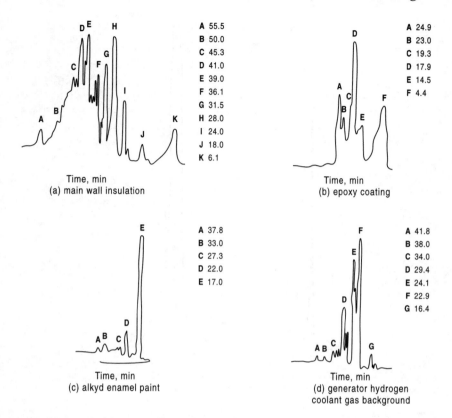

A	55.5
B	50.0
C	45.3
D	41.0
E	39.0
F	36.1
G	31.5
H	28.0
I	24.0
J	18.0
K	6.1

Time, min
(a) main wall insulation

A	24.9
B	23.0
C	19.3
D	17.9
E	14.5
F	4.4

Time, min
(b) epoxy coating

A	37.8
B	33.0
C	27.3
D	22.0
E	17.0

Time, min
(c) alkyd enamel paint

A	41.8
B	38.0
C	34.0
D	29.4
E	24.1
F	22.9
G	16.4

Time, min
(d) generator hydrogen
coolant gas background

Figure 7.7   Comparison of gas chromatograms taken from pre-column samples from hydrogen-cooled generators with overheating of various insulation components. [Taken from Dear et al. [8] © IEEE (1979)]

from pyrolisation is shown in Figure 7.7. Some have used a mass spectrometer in association with the gas chromatograph on the particulate matter, to obtain precise identification of individual compounds. The problem then arises of how to distinguish between the very large numbers of organic compounds that are obtained and how to associate a pattern of compounds with the overheating of a particular insulation material and as yet this has not proved practicable. We shall return to this in the analysis of the gas released in pyrolisation.

An alternative is to reduce the amount of chemical information obtained from the pyrolysis products by using detection techniques that are less sensitive or only sensitive to pyrolysis products. One technique makes use of the fact that many organic materials fluoresce when irradiated with ultra-violet (UV) light. The resultant UV spectrum is far less complex than the chromatogram produced by the same compounds going through a gas chromatograph. Figure 7.8 gives an example taken from Ryder *et al.* [6], which should be compared with Figure 7.6. The filter is illuminated by a

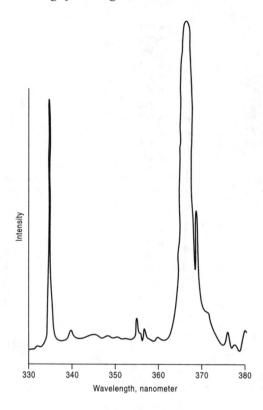

Figure 7.8   Typical ultraviolet spectrum of hydrogen impurities due to pyrolised insulation. [Taken from Ryder et al. [6]. © IEEE (1979)]

UV lamp of a given wavelength, and the fluorescent light from the collected organic particles can be viewed with a UV spectrometer. Although it has been claimed that pyrolised insulation can be clearly distinguished from oil by this technique, to date a commercial version is not available.

It must be stressed that despite the various techniques described here, to the authors' knowledge there is as yet no definitive way to identify conclusively material collected on a core monitor filter. A way out of this difficulty is being sought by tagging components in the machine with compounds, which when overheated give off material with easily identifiable chemical compositions. This technique has been used in the US by Carson *et al.* [2].

7.4.3   Gas analysis off-line

An alternative to detecting and analysing the particulate matter is to detect the gaseous products of pyrolysis, such as the hydrocarbon gases or carbon monoxide and carbon dioxide in the cooling gas. There may be two advantages in doing this. First from

Section 7.2 it is clear that some gases are given off at lower temperatures, before particulates, so an earlier warning of overheating may be given. Second, in a closed cooling circuit, where particulates have a short lifetime of only 15–30 min, gaseous products will have a much longer lifetime of many days, depending upon the coolant gas leakage factor, $1/\tau_r$, so it will not be necessary to detect simply a short-term change. On the other hand, whereas the concentration of particulates will be zero in the absence of burning insulation, there will always be a small and possibly variable background concentration of gases due to impurities in the make-up, and this will effectively determine the threshold for the detection of burning by this technique.

The gas analysis method, which has received attention for large hydrogen-cooled generators, uses a gas chromatograph and flame ionisation detector to measure the total hydrocarbon content in the hydrogen. Early work showed that most organic compounds present in the cooling gas could not be detected without a concentration technique. A number have been reported but the most widely used is a pre-column method in which the impurities are absorbed on a gas chromatographic stationary phase contained in the short tube.

Another off-line method of gas detection that has been used in modern, large, air-cooled machines is to measure the ozone concentration to detect the onset and progress of slot discharge erosion caused by partial discharge activity. Large epoxy insulated windings in an air-cooled environment experience this problem, which can lead to major insulation failure (see Chapters 2 and 10). The technique has been successfully used on large hydro generators, which operate at stator voltages as high as 24 kV in an air-cooled environment, where stator winding bars are subjected to high slot forces and partial discharge activity is consequent upon mechanical damage, leading to the production of ozone gas. The ozone concentration in the coolant air can be measured by taking a gas sample and using a Draeger tube.

7.4.4 Gas analysis on-line

The advantage of performing gas analysis continuously on-line is that, because of the long residence times of gases due to overheating in the cooling system, it may be possible to obtain earlier warning of incipient damage to the machine. The disadvantages are the inherent complexity of continuous chemical gas analysis equipment and the difficulty of translating the analysis into a single electrical signal.

Bearing in mind the experience of Kelley *et al.* [9], shown in Figure 7.9, a continuous monitor was devised for application to hydrogen-cooled generators, using a flame ionisation detector to measure the total organic content of the hydrogen. This is the type of detector used in chromatography for the detection of organic species. The generator hydrogen gas is introduced into a hydrogen/air flame. The flame forms part of an electrical circuit and normally presents a very high resistance. When organic species are introduced, organic ions containing carbon are formed and the resistance of the flame decreases linearly with the amount of organic compound introduced. The device is very sensitive and can detect increases as small as 0.2 parts per million by volume (vpm). However, its usable sensitivity is reduced because of the presence of background levels of organic compounds that can be 10–50 vpm with a variability

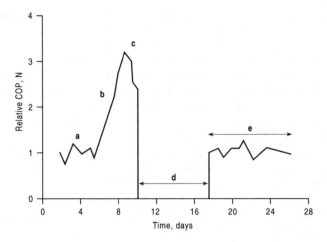

a 1N is equivalent to normal concentration of organic product (COP)
b Significant increase in COP accompanied by a core monitor deflection
c Core monitor alarm, load reduction initiated
d Unit off line for inspection and maintenance
e Generator synchronized, COP back to normal

Figure 7.9   Plot of concentration of organic products during and after a generator
overheating problem. [Taken from Kelley et al. [9]. © IEEE (1976)]

of ±20 per cent. However, one considerable advantage of the continuous monitor over the core monitor is that it shows the trend of any increase in the products of overheating, as shown in Figure 7.9. The total organic content is measured in vpm of methane (CH_4) equivalent. However, Figure 7.10 also shows the very considerable background level against which faults need to be detected. The sensitivity of the monitor has been calculated theoretically based upon the methane equivalent content of typical insulation materials.

This shows that the pyrolysing of 24 g of insulation would produce about 32 1 of methane-equivalent hydrocarbons, which on a large hydrogen-cooled generator (500 MW) would give a concentration of about 15 vpm of methane equivalent on the flame ionisation detector. The amount of hydrocarbons produced depends not only on the temperature of the insulation but also on the part being overheated. The mass of insulation per unit volume in the laminated core is relatively small, whereas on a winding if overheating takes place, a very large proportion of the volume involved will be insulation and this will give a correspondingly large indication on the flame ionisation detector. Overheating would be considered to be serious if the rate of increase of total organics exceeded 20 vpm/h.

An alternative to the flame ionisation detector, for the detection of organic content in hydrogen, has been proposed using a commercial photo-ionisation detector for flammable gases. The detector contains an ultraviolet lamp that ionises the gas

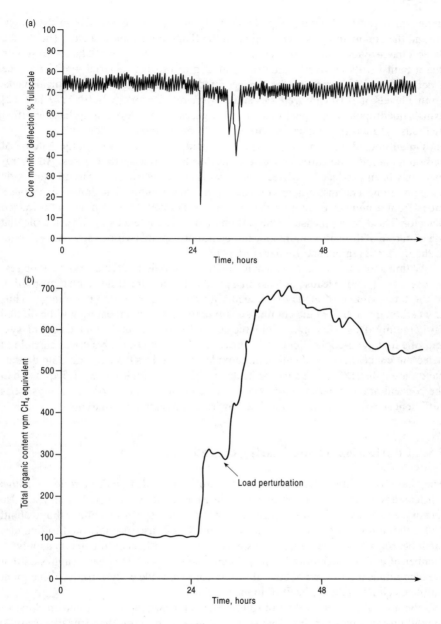

Figure 7.10 *Comparison of the traces from a continuous gas monitor with the response of a core monitor on a 500 MW alternator following an overheating incident. (a) Core monitor response. (b) Continuous gas response.*

stream as it passes through. A potential is applied across electrodes in the detector and the conductivity is measured as in the flame ionisation detector. The device detects the heavier hydrocarbon compounds in the gas stream and it has been shown that a fault involving overheating of about 2 g of organic material would produce a deflection of 1–2 vpm on the device fitted to a large generator. Typical background levels of 7–10 vpm were measured. More recent work by Sorita *et al.* [10–12] using a quadrupole mass-spectrometer on the gases evolved from faults has identified methods for analysing and classifying faults from turbine generator faults.

On air-cooled machines an overheating incident will produce a large volume of carbon monoxide and carbon dioxide as well as light hydrocarbon gases. An instrument has been produced to detect overheating by measuring the carbon monoxide concentration. The instrument contains an auxiliary pump that draws air through tubes from a number of motors that are being sampled, to a commercial infra-red detector. The detector measures the carbon monoxide content using the principle that the vibration of the carbon monoxide molecule corresponds with a known wavelength in the infra-red region, see Tavner *et al.* [4].

Within a totally sealed motor enclosure the air should recirculate with a long residence time, τ_r, but because of leaks in a practical enclosure, a substantial proportion of the total volume of air is exchanged with the environment every minute. Thus any carbon monoxide in the cooling stream produced by overheating will be diluted fairly rapidly. This is in contrast with the detection problem in the sealed cooling system of a hydrogen-cooled generator. However, the infra-red analyser was capable of detecting concentrations of carbon monoxide down to less than 1 vpm. Calculations have shown that 180 g of insulation heated to 300 °C will introduce a 1.5 vpm rise in the concentration of carbon monoxide in the cooling air. Therefore the analyser has sufficient sensitivity to detect localised overheating on motor windings.

7.5 Lubrication oil and bearing degradation

The shafts of smaller electrical machines are supported by ball or roller bearings lubricated with grease, and faults in such bearings could be detected by shock pulse techniques as described in Chapter 8. However, high-speed machines above about 300 kW and low-speed machines above about 50 kW use oil-lubricated rolling element bearings and larger sizes need sleeve bearings with a continuous oil supply. A number of authors including Evans [13] and Bowen *et al.* [14] have suggested that the continuous monitoring of that oil supply could provide early warning of incipient problems either in the oil itself or in the bearings.

The normal mode of failure of rolling element bearings is by fatigue cracking of the rolling elements or their raceways, although other wear mechanisms like fretting, scuffing and abrasion will also generate debris. White-metalled sleeve bearing failures are not usually progressive. Debris is likely to be released in short bursts when the bearing is transiently overloaded or if an oil film momentarily ruptures. Quite substantial damage can be tolerated while the bearing continues to be fed a copious supply of cooled lubricating oil.Nevertheless there are many potentially damaging situations that could be diagnosed by analysing the lubricating oil including fatigue failure or cavitation at the white metal surfaces and corrosion in the lubrication system. Any incipient bearing failure is likely to lead to local heating and degradation of the lubrication oil at the wear site.

A particular problem associated with electrical machines is the flow of currents through bearings and oil films, which pit the bearing surface, producing metallic debris and degrading the lubrication oil. This sort of activity is caused by magnetically induced shaft voltages induced within the machine, whose causes and effects are summarised very thoroughly by Verma and Girgis [15]. Photographs of this type of damage are shown in a previous chapter (Figure 2.14). The two approaches that could be used to detect these various types of incipient failure activity in the lubrication oil are therefore (1) the detection of oil degradation products, and (2) the detection of bearing degradation products or wear debris detection.

7.6 Oil degradation detection

The chemical detection of oil degradation has been used most effectively for the condition monitoring of transformers [16], where the oil is used for insulating and cooling purposes and is sealed within the transformer enclosure. However, in this case the mechanisms of degradation are clearly defined, being distinguished by the different temperatures reached. The technique is off-line gas-chromatograph analysis of the transformer oil, done at regular intervals during the plant life. Such rigorous analysis is necessary to distinguish between the complex products produced in the oil. The mechanisms involved in bearing oil degradation are not so clearly defined and the analysis that would be necessary would be more complex and less easily prescribed. However, standard off-line oil analysis procedures are available commercially but the authors do not know of any programmes that have experienced particular success with electrical machines. The typical parameters to be screened in bearing oil analysis are given in the Neale Report [17].

7.7 Wear debris detection

7.7.1 General

The off-line monitoring of wear debris in lubricating oils has been widely used for some time, particularly for rolling element bearings and gears, in the military and aviation fields. This has made use of the ferromagnetic attraction of debris particles in particular. In fact, passive magnetic plugs fitted to lubricating oil sumps have been used for many years to collect ferromagnetic debris and regular inspection gives an aid to indicate when full maintenance should be carried out on helicopter gearboxes. These techniques have now advanced so that on-line detection of the debris content in oil is possible. Work has also been done to extend the capabilities of on-line detection so that other debris, produced for example from the soft material of white-metal bearings, can be detected. These techniques are described in the following two sections. When reading these sections it should be borne in mind that oil is supplied to the bearing from a closed-loop lubrication system, which will contain oil filtration equipment. This equipment will remove a proportion of any debris entrained so that, following a wear incident, the concentration of debris will increase and then decay away as the filters do their work. Thus the ability to detect wear debris is dependent upon volume and residence time factors similar to those that control the detectability of chemicals in cooling gas, as described in Section 7.3.

7.7.2   Ferromagnetic techniques

The normal mode of failure of rolling element bearings is by fatigue cracking and spalling of the rolling elements or raceways producing fragments up to a millimetre or more in size in the oil. An on-line device has been developed that can be inserted into the full-flow oil line from a bearing and count the number of ferromagnetic particles present in the oil flow, with a certain size band. The detector is based on the induction unbalance principle used in metal detectors. A pair of carefully screened coils surround the oil line and form two arms of an AC bridge circuit. Magnetic or conducting particles entrained in the flowing oil cause the bridge to unbalance, first on one side of null, as the particle approaches the device, and second on the other side of the null as the particle recedes. Figure 7.11 shows a section through the sensor as described in Chapter 4 and by Whittington *et al.* [18]. The phase of the bridge unbalance enables ferromagnetic particles to be discriminated from other conducting particles. The sensitivity of the system varies according to the shape of the particles but for approximately spherical particles the sensitivity can be adjusted to separately record the passage of particles at two size levels, from 200 μm to 2 mm in diameter, in an oil flow velocity of 1–12 m/s; that is, an oil flow of up to 20 l/s. The output is in the form of a counter reading in each of the two size ranges and the output could be made available to a data acquisition system. This robust device has been widely used on jet engine installations, where its performance in a high temperature, pressure and vibration environment has been proved. There is no record, however, of this device being used in an electrical machine.

A device that can produce a greater amount of information about ferromagnetic wear debris is the instrumented magnetic drain plug. Conventional magnetic plugs usually consist of a bar magnet with a pole projecting into the lubricating oil. In an instrumented plug it is necessary to measure the change in field strength at the magnet pole as debris is collected. Because of the difficulties of measuring those changes a horseshoe magnet has been adopted instead of the bar magnet. As particles are attracted to the gap between the poles of the magnet they increase the field within the gap, while at the same time increasing the total flux in the magnetic circuit. Two differentially connected matched field sensors, sampling each of these fields, can thus give additive, particle-dependent signals, while cancelling out fluctuations due to temperature and magnetic field strength. Figure 7.12 shows the arrangement of the device. Another device, an instrumented oil drain plug, gives an analogue voltage proportional to the amount of debris deposited, the rate of accumulation of debris and the temperature of the oil. The device can detect masses of ferromagnetic debris attracted to the pole pieces ranging from 10 to 600 mg with a resolution of 10 mg in an oil flow velocity of 0.1–0.5 m/s.

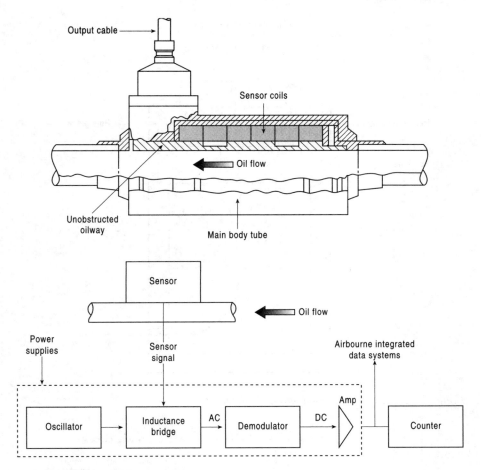

Figure 7.11   Structure of an inductive debris detector

7.7.3   Other wear debris detection techniques

The ferromagnetic techniques described in the previous section are appropriate for rolling element bearings but not for the white-metalled sleeve bearings that are used on larger electrical machines. Lloyd and Cox [19] have investigated the problem of detecting wear in the bearings and hydrogen seals of large turbine generator sets. They describe an investigation that characterised the debris circulating in the oil system of a number of 60 MW turbine generators and correlated the results with plant condition. A major feature in this correlation was the presence of white metal in the machine bearings that typically contain 85 per cent tin.

The results of their investigation showed that by determining the ratio of tin to iron in the debris an operator could see how much bearing damage was occurring, compared to normal running wear (see Figure 7.13). However, if information

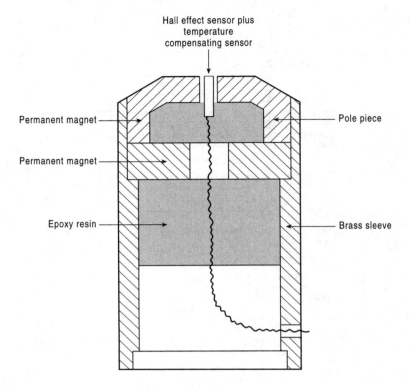

Figure 7.12   An instrumented magnetic drain plug

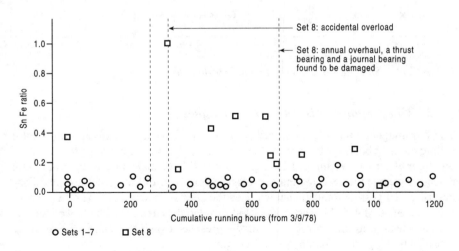

Figure 7.13   Ratio of tin to iron in the oil filter deposits from eight 60 MW turbine generator sets. [Taken from Lloyd and Cox [19]]

was available about running time, monitoring of tin content alone would be adequate.

Lloyd and Cox then proceeded to investigate how the oil system could be automatically monitored to provide early warning of bearing damage. In particular they considered x-ray fluorescence detection and the measurement of the electrical properties of the oil. Their investigation showed that x-ray fluorescence would be feasible but would be prohibitively expensive as an on-line technique.

7.8 Conclusion

This chapter has shown that chemical and wear analysis have been demonstrated to be effective global monitoring techniques for electrical machines which can produce narrow bandwidth (<1 Hz) signals. Chemical degradation of insulating materials and lubricants are detectable and can give bulk indications of the condition of an electrical machine. This is particularly important when it is considered how central insulation and lubrication integrity are to the long-term life of a machine. However, the detectability criteria for these techniques are difficult, the chemical analysis processes involved are complex and expensive, and the quantity of data generated by chemical analysis currently confine their application to only the largest machines.

7.9 References

1. Skala G.F. (1966). The ion chamber detector as a monitor of thermally produced particulates. *Journal de Research Atmospherique* 1966; April/Sept.
2. Carson C.C., Barton S.C. and Echeverria F.S. Immediate warning of local overheating in electrical machines by the detection of pyrolysis products. *IEEE Transactions on Power Apparatus and Systems* 1973; **PAS-92**: 533–42.
3. Tavner, P.J. and Anderson A.F. Core faults in large generators. *IEE Proceedings, Electric Power Applications* 2005; **152**: 1427–39.
4. Tavner, P.J., Gaydon B.G. and Ward D.M. Monitoring generators and large motors. *IEE Proceedings, Part B, Electric Power Applications* 1986; **133**: 169–80.
5. Carson C.C., Barton S.C. and Gill R.S. The occurrence and control of interference from oil-mist in the detection of overheating in a generator. *IEEE Transactions on Power Apparatus and Systems* 1978; **PAS-57**: 1590–614.
6. Ryder D.M., Wood J.W. and Gallagher P.L. The detection and identification of overheated insulation in turbo-generators. *IEEE Transactions on Power Apparatus and Systems* 1979; **PAS-98**: 333–36.
7. Braun J.M. and Brown G. Operational performance of generator condition monitors: comparison of heated and unheated ion chambers. *IEEE Transactions on Energy Conversion* 1990; **EC-5**: 344–49.
8. Dear D.J.A., Dillon A.F. and Freedman A.N. Determination of organic compounds in the hydrogen used for cooling large electricity generators. *Journal of Chromatography* 1977; **137**: 315–22.

9. Kelley J.K., Auld J.W., Herter V.J., Hutchinson, K.A. and Rugenstein, W.A. Early detection and diagnosis of overheating problems in turbine generators by instrumental chemical analysis. *IEEE Transactions on Power Apparatus and Systems* 1976; **PAS-95**: 879–86.
10. Sorita T., Minami S., Adachi H., Takashima N. and Numata S. The detection of degraded materials in turbine generators by chemical analysis. *Proceedings of International Symposium on Electrical Insulating Materials*, Toyohashi, Sep 1998. Piscataway: IEEE; 1998. pp. 751–754.
11. Sorita T., Enmanji K., Kato K., Takashima M. and Nakamura K. On-line detection of overheating material in turbine generators using chemical analysis. *Proceedings of the Annual Conference on Electrical Insulation and Dielectric Phenomena*, Austin, Oct 1999. Piscataway: IEEE; 1999. **2**: pp. 533–6.
12. Sorita T., Enmanji K., Kato K., Takashima M. and Nakamura K. A novel on-line method and equipment to detect local problems in turbine generators. *Proceedings of the Annual Report Conference on Electrical Insulation and Dielectric Phenomena*, Victoria, Oct 2000. Piscataway: IEEE; 2000. pp. 552–5.
13. Evans C. Wear debris analysis and condition monitoring. *NDT International* 1978; **11**: 132–34.
14. Bowen R., Scott D, Seifert WW and Westcott VC. Ferrography. *Tribology International* 1976; **9**: 109–15.
15. Verma S.P. and Girgis R.S. *Shaft Potentials and Currents in Large Turbogenerators*. Report for the Canadian Electrical Association, 078 G 69 1981.
16. Rogers R. Concepts used in the development of the IEEE and IEC codes for the interpretation of incipient faults in power transformers by dissolved gas in oil analysis. Presented at the IEEE Power Engineering Society Winter Power Meeting, New York, 1978.
17. Neale N. and Associates. *A Guide to the Condition Monitoring of Machinery*. London: Her Majesty's Stationary Office; 1979.
18. Whittington H.W., Flynn B.W. and Mills G.H. An on-line wear debris monitor. *Measurement Science Technology* 1992; **3**: 656–61.
19. Lloyd O. and Cox A.F. Monitoring debris in turbine generator oil. *Wear* 1981; **71**: 79–91.

Chapter 8
Vibration monitoring

8.1 Introduction

An electrical machine, its support structure and the load to which it is coupled, form a complex electromechanical system. It can receive impulsive excitation that vibrates it at its own natural frequency, or it can be forced by the exciting airgap electromagnetic field or torque spectrum of the driven or driving machine at many different frequencies. These frequencies may cause the machine to emit an unacceptably high level of acoustic noise, or cause progressive mechanical damage due to high cycle fatigue, which ends in a machine failure mode. Consequently a great deal of effort has been applied to try to determine the principal sources of vibration in electrical machines, and a large literature spanning more than 80 years has accumulated. A representative selection of papers and articles is included in the references.

The principal sources of vibration in electrical machines are:

- the response of the stator core to the attractive force developed magnetically between rotor and stator;
- the dynamic behaviour of the rotor in the bearings as the machines rotates;
- the response of the shaft bearings, supported by the machine structure and foundations, to vibration transmitted from the rotor;
- the response of the stator end windings to the electromagnetic forces on the conductors.

These four areas are interrelated; for example, bearing misalignment or wear can result in eccentric running, which will in turn stimulate the vibration modes of the stator and the rotor. Before examining specific monitoring techniques we shall first look at the characteristic features of the frequency responses of the machine elements listed.

8.2 Stator core response

8.2.1 General

The stator and its support structure comprise a thick-walled cylinder, slotted at the bore, resting inside a thin-walled structure, which may or may not be cylindrical. A stator in its cylindrical support structure is illustrated schematically in Figure 8.1. Analysis of a three-dimensional structure such as this, in response to the complex

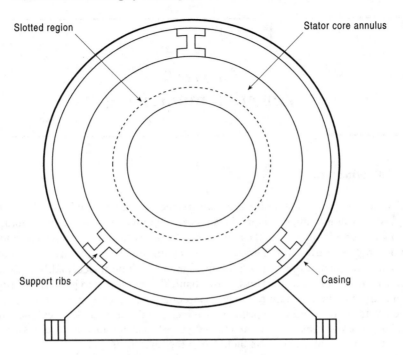

Figure 8.1   A machine stator and its supporting structure

imposed force distribution, is a formidable problem even for powerful numerical techniques. Fortunately the main features of the response can be developed using simplified models.

Alger [1] and Jordan [2] considered the machine to be reducible to an infinitely long, thin-walled cylinder representing the stator core. This simplification allows a qualitative assessment of response to be made. A fuller treatment of the mechanical system is provided by Erdelyi and Erie [3], and Yang [4] gives a detailed analysis of the calculation of stator displacements, taking into account the higher frequency effects due to slot passing.

The forces acting on the stator core are the result of the interaction between the airgap flux wave and the currents flowing in the windings embedded in the stator slots. The forces acting on the end winding are due to the interaction between the end leakage flux and the winding currents. It is apparent, therefore, that the precise nature of the applied force waves will be a function of the form of the current distribution, and the geometry of the airgap and end regions. Disturbances to either, due to rotor eccentricity or damaged areas of the rotor or stator windings for example, will alter the harmonic components of the force wave and initiate a different response from the stator core, particularly if the applied forces stimulate any of the natural modes of the system.

8.2.2   Calculation of natural modes

For a simple thin-walled cylinder of infinite length the radial and peripheral displacements, $u_r(\theta, t)$ and $u_\theta(\theta, t)$, of the structure at a radius fixed at the mid-wall value, may be written as

$$u_r(\theta, t) = \sum_{n=\text{even}}^{\infty} \{a_n(t) \cos(n\theta) + b_n(t) \sin(n\theta)\} \tag{8.1}$$

and

$$u_\theta(\theta, t) = \sum_{n=\text{even}}^{\infty} \{c_n(t) \cos(n\theta) + d_n(t) \sin(n\theta)\} \tag{8.2}$$

If it is further assumed that the deformations are inextensible:

$$\frac{\partial u_\theta}{\partial \theta} = -u_r \tag{8.3}$$

then

$$c_n = \frac{b_n}{n} \text{ and } d_n = -\frac{a_n}{n}$$

During deformation by the deflections in (8.1) and (8.2), the system accumulates volumetric elastic strain potential energy, V_ℓ, per unit length given by the expression

$$V_\ell = \frac{1}{2} \int_{-\pi}^{\pi} \frac{EJ}{(1-v)^2 r^2} \left[\frac{\partial^2 u_r}{\partial \theta^2} - u_r \right] d\theta \tag{8.4}$$

where E is Young's modulus of the material, v is Poisson's ratio of the material and J is the polar moment of inertia of the core cylinder.

Equation (8.4) reduces to

$$V_\ell = \frac{\pi}{2} \frac{EJ}{(1-v^2)r} \sum_{n=1}^{\infty} (1-n^2)^2 \left[a_n^2(t) + b_n{}^2(t) \right] \tag{8.5}$$

Timoshenko [5] showed that the kinetic energy of the system, T_ℓ, per unit length will be given by

$$T_\ell = \frac{w}{2g} \int_{-\pi}^{\pi} (\dot{u}_r^2 + \dot{u}_\theta^2) d\theta \tag{8.6}$$

where w is the weight per unit length per unit circumferential angle of the cylinder and g is the acceleration due to gravity. Substituting (8.1) and (8.2) into (8.6) gives

$$T_\ell = \frac{\pi w}{2g} \sum_{n-1}^{\infty} \left(1 + \frac{1}{n^2} \right) [\dot{a}_n^2(t) + \dot{b}_n^2(t)] \tag{8.7}$$

If it is necessary to include the vibration of the core, enclosure and frame building bars of the machine, then these can also be incorporated. So for example the total elastic strain energy, V_ℓ, would be given by:

$$V_\ell = V_E + V_S + V_F \tag{8.8}$$

where the subscripts E, S and F refer to the enclosure, the stator core and the frame building bars respectively.

The total kinetic energy of the system would be given by

$$T_\ell = T_E + T_S + T_F \tag{8.9}$$

However, the calculation of these quantities for the enclosure and frame structures will be complex. The equation of motion can now be formulated. It is given for free vibration by the Euler–Lagrange equation:

$$\frac{\partial}{\partial t} \left(\frac{\partial T_\ell}{\partial \dot{a}_n} \right) - \frac{\partial V_\ell}{\partial \dot{a}_n} = 0 \tag{8.10}$$

If the forced vibration response is required directly then the right hand side of (8.10) becomes a forcing function. For example, if we are interested in the system response to the mth harmonic of the radial force wave $f(\theta, t)$, the forcing function becomes the work function, W, given by

$$W = \sum_{m=1}^{\infty} f_m(\theta, t) u_{rm}(\theta, t) \tag{8.11}$$

Generally, $f_m(\theta, t)$ has the form of travelling wave:

$$f_m(\theta, t) = F_m \cos(\omega_m t - m\theta) \tag{8.12}$$

Erdelyi and Erie [3] shows how the Rayleigh–Ritz method can be used to solve (8.10) to yield the natural frequencies of the system, under the assumption that time variations are harmonic. That is, the coefficients $a_n(t)$ and $b_n(t)$ are given by:

$$a_n(t) = A_n \sin \omega_n t \text{ and } b_n(t) = B_n \sin \omega_n t \tag{8.13}$$

Under this assumption, (8.10) becomes a polynomial in ω_n whose roots give the natural frequencies. Some of the natural circumferential mode shapes of the radial vibration of the stator are shown in Figure 8.2.

Besides these circumferential modes it is also possible for the radial vibration of the stator to vary as a function of machine length. The mode shapes for a cylinder vibrating in this way are illustrated in Figure 8.3. In practice, however, only the $k_L = 0$ case need be considered for machines of normal proportion. Very long machines may exhibit vibrations at higher modes, but the most important mode shapes are those due to the circumferential radial vibrations. It is easily recognised that the cases for $k_c = 0$ and $k_L = 0$ and are identical.

Approximate formulae for the natural frequencies of a simple single ring stator have been reported by Yang [4] in the form given here. For $k_c = 0$, the corresponding

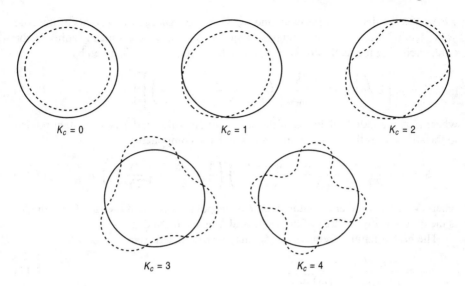

Figure 8.2   Radial mode shapes of a stator in a circumferential direction

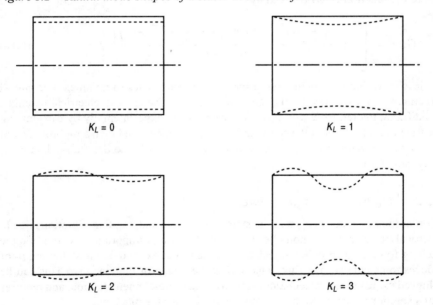

Figure 8.3   Radial mode shapes of a stator in a longitudinal direction

natural frequency f_0 is given by

$$f_0 = \frac{1}{2\pi \, r_{\text{mean}}} \left[\frac{E}{\rho} \left(\frac{w_y}{w_y + w_t + w_i + w_w} \right) \right]^{1/2} \tag{8.14}$$

where ρ is the density of the core, and w_y, w_t, w_i and w_w are the weights of the core yoke, teeth, insulation and windings, respectively, r_{mean} is the mean radius of the core, excluding the teeth. For $k_c = 1$, the natural frequency f_1 is given by

$$f_1 = f_0 \left[2 \bigg/ \left\{ 1 + \frac{t_0^2 S}{12\, r_{mean}^2} \left(\frac{w_y}{w_y + w_t + w_i + w_w} \right) \right\} \right]^{1/2} \tag{8.15}$$

where t_0 is the radial thickness of the stator core annulus and S is a constant related to the stiffness of the winding, insulation and slot components:

$$S = 1 + \frac{N_s}{2\pi\, J\, r_{mean}} \left(\frac{w_t + w_i + w_w}{w_t} \right) \left(\frac{1}{3} + \frac{t_0}{2h_t} + \left(\frac{t_0}{2h_t} \right)^2 \right) a_t h_t^3 \tag{8.16}$$

where N_s is the number of stator slots, J is the polar moment of inertia of the core, a_t is the cross-section of area of the teeth, and h_t is the tooth depth.

The higher natural frequencies, f_m, may be calculated from

$$f_m = \frac{f_0 t_0 m (m^2 - 1) G(m)}{(2/3) r_{mean} \sqrt{(m^2 + 1)}} \tag{8.17}$$

where the function $G(m)$ is given by

$$G(m) = \left[1 + \frac{\frac{t_0^2}{12 r_{mean}^2} (m^2 - 1) \left\{ m^2 \left(4 + \frac{w_y M}{w_y + w_t + w_i + w_w} \right) \right\}}{(m^2 + 1)} \right]^{-1/2} \tag{8.18}$$

It is clear that the calculation of the natural frequencies of complex mechanical structures, as represented by the the stator core and frame of an electrical machine, is a difficult matter. However, this could be resolved experimentally by carrying out separate mobility tests of the stator and rotor components and then a mobility test of the composite assembly. By these means it will be possible to determine the modes of the structure.

8.2.3   Stator electromagnetic force wave

In order to anticipate changes in the stator core frame and winding vibrations due to electrical or mechanical anomalies in the machine, it is important to determine the exciting forces. The problem of calculating the forces exerted on the stator and rotor reduces to calculating the flux density $\bar{B}$ in the airgap of the machine. This can be achieved using numerical techniques, such as the finite-element method, and requires that a solution be found to the following equation for the machine:

$$\bar{\nabla} \times \bar{\nabla} \times \bar{B} = \sigma \left(\bar{v} \times \bar{B} - \mu \frac{\partial \bar{B}}{\partial t} \right) \tag{8.19}$$

where σ is the electrical conductivity of the region, and v is the velocity of the rotor, relative to the travelling flux wave produced by the stator. The solution of (8.19), even in the two dimensions representing a cross-section in the radial and circumferential directions of the machine, requires significant computational effort if the slotting of both rotor and stator are to be taken into account. If good accuracy is required, however, it is the only suitable path to follow. Penman *et al.* [6] gave a detailed explanation of the solution of (8.19) for a linear machine, using the finite-element technique.

The finite-element method can be used to account for any distribution of windings and any radial or peripheral geometrical variations, simply by increasing the complexity of the model. However, there are limits to the accuracy of finite-element modelling determined by the accuracy of the precise manufacturing data. These methods are essentially numerical and therefore quantitative. A qualitative assessment is often more valuable than a full analysis and this is readily achieved, in certain circumstances, using simpler (although less accurate) methods.

If the rotor and stator surfaces are assumed to be smooth then it is possible to solve (8.19) in the radial and circumferential plane analytically, provided the motion term is neglected. This approach allows the effect of individual conductor currents to be accounted for. Hague [7] and Stafl [8], both used separation of variable techniques to calculate airgap flux densities for a variety of configurations.

Many authors have examined the important sources of unbalanced magnetic pull (UMP) and the effect it has on vibration. In particular Binns and Dye [9] identify the role of static eccentricity in the production of UMP. In addition, Swann [10] shows how it is possible to calculate the harmonics introduced into the flux wave due to rotor eccentricity. He does this by using a conformal transformation to re-centre the rotor. A review of a considerable amount of previous work on UMP may be found in the report by Rai [11].

Perhaps the simplest, generally useful method of gaining a qualitative assessment of the flux waveform is that used by Yang [4]. Here the flux wave is calculated by simply multiplying the stator magnetomotive force distribution (F_1) due to the winding currents, by the permeance (Λ) of the airgap:

$$\Phi = F_1 \Lambda \tag{8.20}$$

Binns [12] suggested that this procedure had limited accuracy, but within the limitations suggested by Lim [13] the technique is valuable, since it can easily accommodate geometrical effects and anomalies in winding arrangements. Williamson and Smith [14] later proposed an analytical permeance-wave method for calculating the harmonic impedances of a machine and, by use of harmonic equivalent circuits, it was therefore possible to predict machine rotor torques. Williamson and other authors have subsequently successfully used this method, sometimes incorporating finite-element solutions, to allow for three-dimensional components that are not tractable by the analytical method. Using such techniques it has been possible to analyse the effects of winding faults and rotor eccentricity on the flux density in the airgap.

Vas made the point in his book [15] about the all-embracing nature of the electromagnetic field in the energy conversion process, stressing the central importance of the airgap flux density in monitoring electrical machines. This point relates to the emphasis there has been to date in the published literature on the effect of the more tractable faults on the airgap field, as mentioned in Chapter 3.

The permeance variation, taking into account the relative motion of the rotor with respect to the stator, can be expressed in the form of an infinite series of harmonics. If both the rotor and stator surfaces are slotted then the permeance wave has the form

$$\Lambda(\theta_1, t) = \sum_{m=1}^{\infty} \sum_{n=1}^{\infty} \hat{\Lambda}_{m,n} \cos[mN_r \omega_{rm} t - (mN_r \pm nN_s)\theta_1] \tag{8.21}$$

with $\hat{\Lambda}_{m,n}$ the peak amplitude of the permeance wave defined above, m and n are integer slot harmonics, N_s and N_r are number of stator and rotor slots respectively, and ω_{rm} is the mechanical angular speed of the rotor.

Similarly the magetomotive force (MMF) of the stator, f_1, can be expressed as an infinite series of space and time harmonics. The result, which can be found in standard texts, is

$$f_1(t) = \sum_{\ell=1}^{\infty} \sum_{q=1}^{\infty} \hat{f}_{s,\ell,q} \cos \left[\ell \omega_{se} t - qp \left(\theta_1 - \frac{\alpha_s z}{L} \right) - \phi_{\ell q} \right] \tag{8.22}$$

where $\hat{f}_{s,\ell,q}$ is the stator MMF wave components of the form NI with ℓ time harmonics and q space harmonics, q is the order of the stator winding MMF space harmonic, ℓ is the order of the supply time harmonics, ω_{se} is the angular frequency of the electrical supply, z is the longitudinal distance from the centre of the machine, L is the active length of the core, α_s is the skew angle of the stator, $\phi_{\ell q}$ is the phase angle of the stator MMF wave $f_{s,\ell,q}$, and p is the number of pole pairs.

Similarly the rotor MMF, referred to the stator, can be expressed as

$$f_2(t) = \sum_{\ell=1}^{\infty} \sum_{q=1}^{\infty} \hat{f}_{r,\ell,q} \cos \left[\ell s \omega_{se} t - kp \left(\theta_2 - \frac{\alpha_r z}{L} \right) - \phi_{\ell q} \right] \tag{8.23}$$

where s is the slip of the rotor with respect to the stator magnetic field, $\hat{f}_{r,\ell,q}$ is the rotor MMF wave components of the form NI with ℓ time harmonics and k space harmonics. The total MMF can be found by adding (8.22) and (8.23), and the flux wave calculated by multiplying this sum by the result of (8.21).

Yang shows that the effect of eccentricity can be incorporated by modifying the expression for the permeance wave (8.21). He proposes the following expressions for dynamic and static eccentricity.

- Stator permeance variations, $\Lambda_{s,ecc}$, due to static eccentricity:

$$\Lambda_{s,ecc} = \sum_{n=1}^{\infty} \Lambda_{0,ecc} \cos n\theta_1 \tag{8.24}$$

- Stator permeance variations, $\Lambda_{d,ecc}$, due to dynamic eccentricity:

$$\Lambda_{d,ecc} = \sum_{n=1}^{\infty} \hat{\Lambda}_{0,ecc} \cos(\omega_{ecc} t - n\theta_1) \tag{8.25}$$

where ω_{ecc} is the angular velocity of the eccentricity.

These expressions can be combined with (8.21) to give the permeance wave for the complete system and the flux wave found in the manner outlined. Ovality of the stator bore can be accounted for in a similar manner. The expressions and methodology outlined here provide a general and relatively simple method of calculating the harmonics of the flux wave acting on the stator so that its response can be determined. The radial and tangential forces applied to the core can be calculated from the flux density using the method of Maxwell stresses, as described by Carpenter [16] from the radial and tangential stresses given by σ_r and σ_θ:

$$\sigma_r = \frac{B_r^2}{2\mu_0} \text{ and } \sigma_\theta = \frac{B_\theta^2}{2\mu_0} \tag{8.26}$$

where B_r and B_θ are the amplitudes of the radial and tangential flux density waves, calculated from the MMF and permeance waves that when integrated give (8.20).

8.3   Stator end-winding response

The end-winding structure of an electrical machine has a relatively low stiffness or compliance but relatively high non-linear damping coefficients due to frictional contact between adjacent conductors in the structure. The stiffness may be increased by improved methods of bracing, which are used in large turbine generators or induction machines with more onerous starting duties. The motion of the stator end winding is excited by two mechanisms:

- seismic excitation of the coils, as encastre beams, by the ovalising displacements of the stator core and displacement of the machine by its environment;
- electromagnetic forces on the coils themselves due to the currents flowing in them.

These latter forces have been considered by Brandl [17]. The dynamics of the end winding are very complex, partly because of its complicated geometry but also because of the distributed nature of the forces applied to it and the non-linear coefficients of its response. The resultant displacements are at twice the electrical supply frequency, f_{se}, and it is necessary to carry out a very thorough analysis to determine the mode shapes of the structures. The dynamic behaviour has been described by Ohtaguro *et al.* [18]. Again when analysis is difficult and experimental determination of the structure can be obtained by mobility tests to determine the modal response of the structure.

A number of utilities have installed triaxial accelerometers on the end-winding structures of large turbine generators to monitor the amplitudes of $2f_{sm}$ vibrations at widely spaced intervals of time in order to check that the end winding has not slackened. Displacements of end windings on large turbine generators of 1–10 mm are quite usual during normal running, even with modern epoxy insulation systems, and this can lead to fretting of the insulation, slackening and dielectric damage.

Because of this, end-winding monitoring is now being used for routine on-line purposes on the very largest synchronous machines but caution must be taken, as it requires very specialist interpretation to determine from accelerometer signals where end-winding slackening has occurred so that remedial action can take place. However, on smaller machines, for example induction motors, the displacements are not so large, but do require special techniques for prediction [19], and for measurement [20], particularly during starting when winding currents are very large.

8.4   Rotor response

We now consider the motion of a rotor in response to:

- transverse force excitation, due to self-weight, mechanical unbalance, shaft whirling, dynamic or static electromagnetic UMP due to eccentricity or a combination of all four;
- torsional torque excitation, due to the prime mover drive or electromagnetic torque reaction.

Transverse forces are due to asymmetries in the machine, while torsion is primarily due to the driving torque; however, both may be affected by electrical or mechanical faults in the machine itself or electrical or mechanical system disturbances outside the machine. An important issue is the response of the machine to these applied excitations. There will also be a coupling between torsional and transverse effects due to the transfer function or stiffness between these axes of the machine, so torsional effects, like current faults in rotor and stator windings, can cause transverse effects like vibrations, and vice versa.

8.4.1   Transverse response

8.4.1.1   Rigid rotors

In order to examine the response of a rotor to unbalanced forces a distinction must be drawn between rigid and flexible rotors. Rigid or short rotors may be considered as a single mass acting at the bearings, and Wort [21] showed that the displacement, u, of the rotor and its bearings can be modelled by the differential following equation:

$$(M + M_s)\frac{d^2u}{dt^2} + c\frac{du}{dt} + ku = mrw_{rm}^2 \tag{8.27}$$

where M is the mass of the rotating system, m is the equivalent unbalance mass on the shaft, M_s is the support system mass, r is the effective radius of the equivalent unbalanced mass, c is the damping constant of the support system, and k is the stiffness of the support system

For sinusoidal motion the peak displacement, $\hat{u}$, is given by the solution to (8.27), and is

$$\hat{u} = \frac{mr\,(\omega_{rm}/\omega_0)^2}{((M + M_s)\sqrt{\{(1 - (\omega_{rm}/\omega_0)^2)^2 + 4D^2(\omega_{rm}/\omega_0)^2\})}} \tag{8.28}$$

with ω_0, the natural frequency of the rotor support system, and D, the damping factor, given by

$$\omega_0 = \sqrt{\frac{k}{(M + M_s)}}$$

$$D = \frac{c}{2\sqrt{k(M + M_s)}}$$

If the displacement is divided by the specific unbalance e, given by

$$e = \frac{mr}{M} \tag{8.29}$$

then the behaviour of the displacement, u, as a function of frequency is as shown in Figure 8.4. The degree of residual unbalance is denoted by the quantity $G = e\omega_{rm}$ and the permissible limits are provided by international standard ISO1940-1:2003 [22] as shown in Figure 8.5.

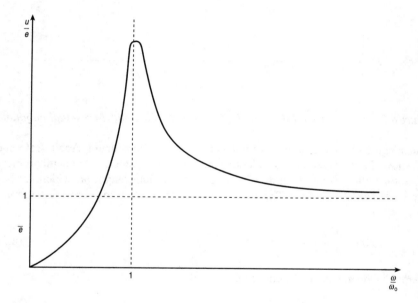

Figure 8.4 Displacement per specific unbalance versus normalised frequency

For electrical machinery the appropriate quality grades are towards the lower end of the values shown in Figure 8.5. For example, $G = 2.5$ is generally applicable to machines of all sizes, and $G = 1.0$ for special requirements.

8.4.1.2 Flexible rotors

For long, slender rotors operating at higher speeds, as in two-pole machines, particularly large turbine generators that have restricted rotor radii, the foregoing analysis is

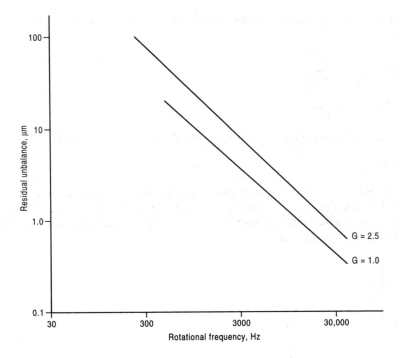

Figure 8.5   Extract from ISO 1940-1:2003 permissible limits to residual unbalance

insufficient and the distribution of unbalance must be considered. Again Reference [21] shows how it is possible to calculate the natural frequencies of general problems of a rotor with a flexural rigidity *EJ*, and mass per unit length *m*, which are both functions of axial location *z*. The displacement *u*, for any *z*, is given by the solution of

$$\frac{\partial^2}{\partial z^2}\left(EJ(z)\frac{\partial^2 u}{\partial z^2}\right) - \omega_{rm}{}^2 m(z)u = \sum_{n=1}^{\infty} f_n \tag{8.30}$$

where f_n is the unbalance forcing function and

$$2\pi f_n = \omega_n = \int m(z)g_n^2(z)dz \tag{8.31}$$

with $g_n(z)$ the nth solution of equation (8.30). The solution for coupled systems comprising several rotors, consisting of the electrical machine and the machine it is driving or driven by, is extremely complex. It is usual to assume that the stiffness of couplings between rotors is low and rotors may be considered to be decoupled from one another, allowing them to be considered individually as described earlier.

The mode shapes for the rotor shafts will also depend upon the nature of the bearing supports for the shafts. For example Figure 8.6 shows the effect of hard and soft bearings on the first and second modes for a single flexible rotor.

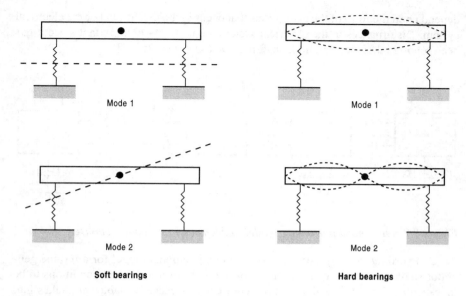

Figure 8.6   Rotor mode shapes for hard and soft bearings

The results for higher modes are readily deduced from these examples. For nearly all electrical machine vibrations, for which monitoring would be appropriate, the bearings may be considered to be hard, although Smart *et al.* [23] have shown how the influence of foundations can be taken into account when studying the transverse vibration behaviour of a large machine.

The international standard limits shown in Figure 8.5 are applicable to rigid rotors operating well below their critical speed. For flexible rotors it has been suggested that the allowable eccentricity can be modified so that the same standard applies. Dimentberg [24] uses the correction

$$e_{flexible} < e_{rigid} \left(1 - \left(\frac{\omega_{rm}}{\omega_0} \right)^2 \right) \tag{8.32}$$

where ω_0 is the first critical speed of the rotor. This, however, is only applicable for rotors running at less than ω_0; that is, a hypocritical machine. When the machine is operating above ω_0 it is a hypercritical machine.

8.4.2   Torsional response

The torsional oscillatory behaviour of a large rotating electrical machine and its driven machine or prime mover can be complicated. For example, in a large turbine generator the electrical machine links a complex prime mover to a large interconnected electrical network in which large quantities of energy are being transported. The possibility of forced torsional oscillation in the rotor of a turbine generator is significant because of its great length and relatively small radius. The nature of such oscillations will depend upon the form of the disruptions that occur in the mechanical or the electrical system. Disturbances in the electrical system are particularly important since it has been reported that they can limit shaft life due to fatigue [25–27].

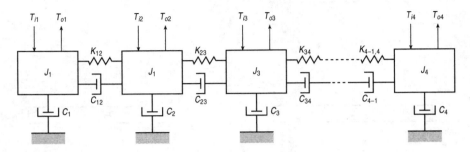

Figure 8.7   A lumped parameter model to determine torsional response

Cudworth *et al.* [28] have developed a computational model for a turbine generator shaft that is very general, and allows a wide variety of fault conditions to be investigated. The electrical system is represented in phase variables and takes into account as many static and rotating electrical elements as may be required. For the mechanical system the lumped parameter model illustrated in Figure 8.7 is used. The shaft damping and steam damping effects on the turbine blades are represented by variable viscous dampers and material damping is also included.

The waveform shown in Figure 8.8 is representative of the results achievable using a model such as the one described earlier. This shows the oscillations in shaft torque

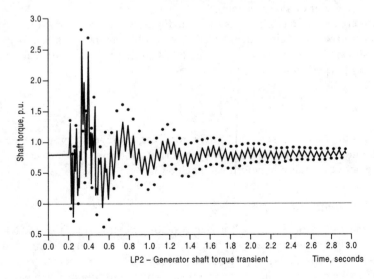

Figure 8.8   Calculated shaft torque transients using a lumped parameter model, the dots showing the permissible limits of error in the calculation

of a 500 MW generator following a phase-to-phase short circuit on the transmission system. The lower level oscillations persist for some considerable time after the major transient has largely died away and this has obvious implications for the fatigue life of the shaft. It may be advisable to monitor such events, and methods for doing so will be discussed later in this chapter.

8.5   Bearing response

8.5.1   General

Rotor vibration force is transmitted to the stator via the airgap magnetic field and the bearings in parallel. It is therefore important to consider the response of the bearings to that vibration force so that its effect is not confused with vibrations generated by faults within the bearings themselves. Rotor vibration force will cause vibration of the rotor relative to the housing and an absolute vibration of the bearing housing. This action must be considered for both rolling element bearings and oil-lubricated sleeve bearings.

8.5.2   Rolling element bearings

A schematic view of a typical rolling element bearing is shown in Figure 8.9 and these are fitted to machines generally below 300 kW in output, which constitute the vast majority of electrical machines.

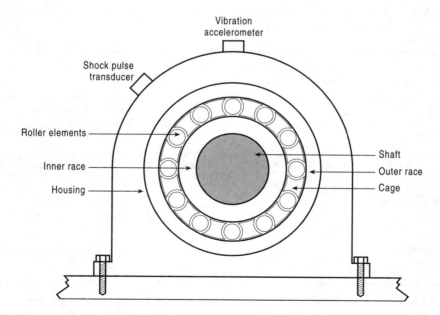

Figure 8.9   Rolling element bearing assembly

The failure of rolling element bearings is the most common failure mode associated with smaller machines. Because of their construction, rolling element bearings generate precisely identifiable vibration frequencies. Also, since the oil or grease film is very thin, the relative motion between the housing and the shaft is small. It is therefore possible to detect on the stator side the vibrations associated with the bearings using an accelerometer mounted directly on the bearing housing, ω_{sm}.

The characteristic frequencies of rolling element bearings depend on the geometrical size of the various elements, and can be found in many texts (see Collacott [29] for example). Table 8.1 summarises these frequencies and their origins.

Besides the frequencies given in Table 8.1, there will also be higher frequencies generated by elastic deformation of the rolling elements themselves, and the excitation of the natural modes of the rings that comprise the inner and outer races. These effects will, however, be secondary to the principal components defined here. The magnitudes of the components given in Table 8.1 are often lost in the general background noise when the degree of damage is small, but because of their precise nature they present an effective route for monitoring progressive bearing degradation. A simple instrument can be devised using an accelerometer mounted on the bearing housing to detect the amplitude of vibration at these characteristic frequencies. Once the characteristic frequencies have been calculated it is possible to enhance the performance of the instrument by the use of highly selective filters and weighting functions, so as to be able to identify bearing faults at an earlier stage.

Table 8.1   *Characteristic transverse mechanical angular frequencies produced by rolling element bearings*

Fault	Stator transverse mechanical frequency, w_{sm}, rad/s	Comment
Outer race	$\omega_{sm} = \dfrac{n_b}{2}\omega_{rm}\left(1 - \dfrac{d_b}{D_b}\cos\phi\right)$	Rolling element passing frequency on the outer race
Inner race	$\omega_{sm} = \dfrac{n_b}{2}\omega_{rm}\left(1 + \dfrac{d_b}{D_b}\cos\phi\right)$	Rolling element passing frequency on the inner race
Rolling element	$\omega_{sm} = \dfrac{D_b}{2d_b}\omega_{rm}\left(1 - \left(\dfrac{d_b}{D_b}\right)^2\cos^2\phi\right)$	Rolling element spin frequency
Train fault	$\omega_{sm} = \dfrac{1}{2}\omega_{rm}\left(1 - \dfrac{d_b}{D_b}\cos\phi\right)$	Caused by an irregularity in the train

Definitions: n_b = number of rolling elements; $\omega_{rm} = \frac{2\pi N}{60}$ mechanical rotational frequency, Hz; w_{sm} = mechanical vibration frequency on the stator side; N = rotational speed in rev/min; d_b = rolling element diameter; D_b = rolling element pitch; f = rolling element contact angle with races.

When monitoring the vibration due to rolling element bearings it is prudent to obtain a good vibration baseline. This is because once the bearing becomes worn significantly the vibration spectrum it emits becomes more random again, although at a higher baseline than for a good bearing. If no baseline is available, no history has been built up and the background noise has risen, then it will be impossible to detect specific faults. Machinery will also have a degree of unbalance, which will modulate the characteristic frequencies of the bearings and produce side bands at the rotational frequency. Vibration monitoring is highly suitable for rolling element bearings, although it is complex but has gained wide acceptance throughout the industry. Stack *et al.* [30] give a description of the process to classify rolling element bearing faults.

8.5.3   Sleeve bearings

In sleeve bearings, the shaft is supported by a fluid film pumped, by the motion of the shaft, at high pressure into the space between the bearing liner and the shaft. Because of the compliance of the oil film and the limited flexibility of the bearing housing, vibrations measured at the housing may be of low amplitude. Also, because the liners of the bearing will inevitably be a soft material such as white metal, small faults are very difficult to identify by measuring the absolute vibration of the housing. These factors point to the use of displacement transducers as being the most effective tool, but they will only be useful at lower frequencies. Higher frequencies; for example multiples of the rotational frequency, are best measured with an accelerometer mounted on the bearing housing. It is worth bearing in mind, however, that as the bearing load increases due to an increase in rotor load, the oil film thickness decreases with a commensurate decrease in bearing flexibility. This increases the vibration detectable at the bearing housing and allows more information to be derived from the measurement.

An important concern for sleeve bearings is the onset of instability in the oil film. This can result in oil whirl and subsequently oil whip, in response to unusual loading of the bearing. Figure 8.10 shows the forces acting upon the shaft in a sleeve bearing, and illustrates that the shaft is supported by a wedge of oil just at the point of minimum clearance.

The oil film is circulating at a speed of approximately half the shaft speed, but because of the pressure difference on either side of the minimum clearance point, the shaft precesses at just below half speed. This motion is termed oil whirl and is a direct result of the pressure difference mentioned earlier, which comes about due to viscous loss in the lubricant. Instabilities occur when the whirl frequency corresponds to the natural frequency of the shaft. Under such conditions the oil film may no longer be able to support the weight of the shaft. Details of the mechanisms involved in oil whirl and its development into the more serious instability called oil whip (which occurs when the shaft speed is twice its natural frequency) are given by Ehrich and Childs [31]. Care must be taken, therefore, that either the machine does not operate at a speed higher than twice the first critical speed of the rotor shaft, or if it must,

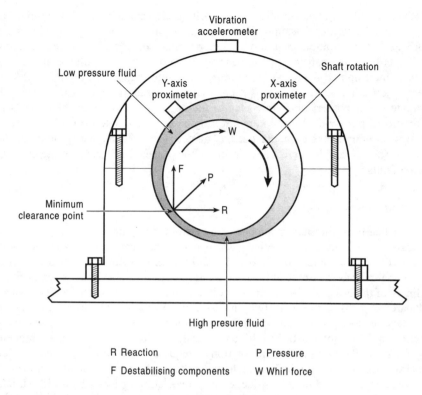

Figure 8.10   Forces acting upon a shaft in a sleeve bearing

then oil whirl must be suppressed. The following frequency, ω_{sm}, is identified in the literature as caused by oil whirl:

$$\omega_{sm} = (0.43 \text{ to } 0.48)\omega_{rm} \tag{8.33}$$

8.6   Monitoring techniques

Now that we have discussed the ways in which transverse and torsional vibration can be produced in electrical machines, and we have outlined, in Chapter 3, the principal analytical tools at our disposal to measure the effect of vibration, we can proceed to show how vibration can be used to monitor the health of machines.

For electrical machine condition monitoring it is important to recognise that mechanical and electrical faults excite the machine structure in different ways. For example:

- Mechanical faults like self-weight, mechanical unbalance and shaft whirling will excite transverse motion in the machine frame, detectable by vibration sensors.

- Dynamic or static electromagnetic UMP due to eccentricity, which may be caused by bearing wear, will also excite transverse motion in the machine frame, detectable by vibration sensors.
- Electrical faults in stator or rotor windings will excite torsional motion in the shaft that will be detectable in the torque signal but not necessarily by vibration sensors, unless the activity is coupled to transverse motion by asymmetries in the machine frame.

Because many faults can be identified by more than one method, and because the internationally agreed standards and limits relate not to specific items of plant, but to the form of analysis to which measurement is subjected, we shall use the measurement treatment as the generic identifier. We shall also be principally concerned with so called on-line monitoring; that is, we are interested in techniques that can be applied to machinery that is running and in the predictive power of the monitoring rather than simply the ability to merely intercept faulty conditions when they become serious enough to cause damage. There may often be some commonality between these objectives.

8.6.1   Overall level monitoring

This simple form of monitoring is the most commonly used technique, although as an aid to diagnosis of faults in electrical machines its efficacy is limited. The measurement taken is simply the rms value of the vibration level on the stator side of the machine over a selected bandwidth. The usual bandwidth is 0.01–1 kHz or 0.01–10 kHz, and in practice the measured parameter is usually vibration velocity taken at the bearing cap of the machine under surveillance. The technique has found favour because over the years a considerable experience has been built up to relate overall vibration monitoring levels to machinery failure modes. This has resulted in the publication of recommended vibration standards for running machinery. These standards do not give diagnostic information, but simply indicate the overall health of machinery at a given vibration level. Many operators use such information to develop a strategy for maintenance scheduling.

The guidance given in the past by the German Vibration Standard VDI 2056 [32] is illustrated in Table 8.2; however, the relevant up-to-date standard is ISO 10816-1:1995 [33]. These criteria are based solely on machine rating and support systems, and utilise a 0.01–1 kHz bandwidth. Essentially it recommends that when vibration levels change by 8 dB or more care must be exercised, and when the change exceeds 20 dB action should follow. These limits can be relaxed, however, if a subsequent frequency analysis shows that the cause of the increase in level is due to a rise in the higher frequency components. In such cases changes of 16 dB and 40 dB respectively may be more appropriate. A section of the specification, as it relates to electrical machines, is given in Table 8.3. Group K signifies smaller, quiet running plant, group M is medium-sized plant and group G is the larger, noisier plant. This reflects that electrical machines, when uncoupled from their prime mover or driven plant, are generally low-noise, low-vibration machines.

Table 8.2   Vibration standard VDI 2056

Vibration velocity, mm/s rms	Vibration velocity, dB, ref 10^{-6} mm/s	Group K	Group M	Group G
45	153	Not permissible	Not permissible	Not permissible
28	149		20 dB ($\times 10$)	
18	145			
11.2	141			Just tolerable
7.1	137		Just tolerable	
4.5	133	Just tolerable		Allowable
2.8	129		Allowable	
1.8	125	Allowable		Good large
1.12	121		Good medium	machines with
0.71	117	Good small	machines	rigid and heavy
0.45	119	machines up	15–75 kW or up	foundations
0.28	109	to 15 kW	to 300 kW on	whose natural
0.18	105		special	frequency
			foundations	exceeds machine speed

Table 8.3   Vibration limits for maintenance as given in CDA/MS/NVSH 107

Type of plant	New machines		Worn machines (full-load operation)	
	100–1 000 hour life, mm/s	1000–10 000 hour life, mm/s	Service level, mm/s	Immediate overhaul*, mm/s
Boiler auxiliaries	1.0	3.2	5.6	10.0
Large steam turbine	1.8	18.0	18.0	32.0
Motor-generator set	1.0	3.2	5.6	10.0
Pump drives	1.4	5.6	10.0	18.0
Fan drives	1.0	3.2	5.6	10.0
Motors in general	0.25	1.8	3.2	5.6

*These levels must not be exceeded in any octave band

Another useful set of criteria was given in the Canadian Government Specification CDA/MS/NVSH107 [34], which is no longer issued. This specification relates primarily to measurements taken on bearings, and it is here that overall level measurement is most commonly employed. This specification has a broader bandwidth than VDI 2056, namely 0.01–10 kHz, but still relies on overall velocity vibration measurement.

The strength of the overall vibration level technique is its simplicity. It requires only the simplest of instrumentation applied to the stator side of the machine, and because of this it is a common feature in many installations. It also provides an ideal method for use with portable instruments, but it makes heavy demands upon technical personnel. The sensitivity of the technique is also low, particularly when a fault is at an early stage, and there is little help on offer to aid diagnosis without further sophisticated techniques being employed.

The Neale Report [35] indicates that it may be possible to effect a limited diagnosis by taking two overall level measurements: V_a, the peak vibration velocity in mm/s; and X, the peak to peak displacement in μm. These quantities are then used to define the parameter F as follows:

$$F = \frac{0.52NX}{V_a} \qquad (8.34)$$

where N is the speed in rev/min of the machine. The interpretations in Table 8.4 are then appropriate.

Table 8.4   *Diagnosis using overall level measurements based on the Neale Report [35]*

Value of F	Trend	Fault
<1	Decreasing	Oil whirl
$=1$	Steady	Unbalance, indicative of eccentricity or perhaps faulty rotor
>1	Increasing	Misalignment, static eccentricity

8.6.2   *Frequency spectrum monitoring*

The key to vibration monitoring diagnostics is the frequency spectrum of the signal, and the past decade has seen a remarkable increase in the range and sophistication of techniques and instrumentation available for spectral analysis. There are various levels of spectral analysis commonly used, and these may be regarded as a continuum extending from the overall level reading to the narrow band with constant frequency bandwidth presentation, as shown in Figure 8.11.

In the octave band and third-octave band techniques the spectrum is split into discrete bands, as defined by Figure 8.12. The bands are, by definition, such that when the frequency is scaled logarithmically the bands are of equal width. The constant percentage band is one which is always the same percentage of the centre frequency, while the constant frequency bandwidth is an absolutely fixed bandwidth form of analysis that can give very high resolution, provided the instrumentation is of a sufficiently high specification.

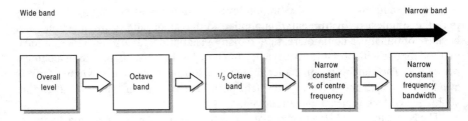

Figure 8.11   Levels of spectral analysis

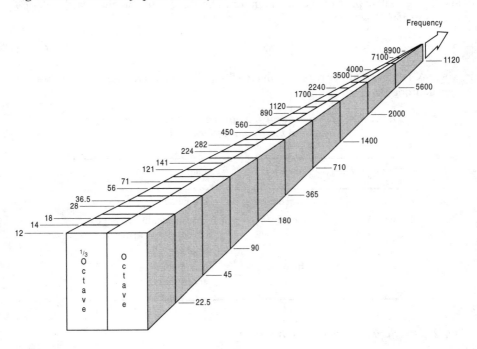

Figure 8.12   Octave and third-octave bands

The effect that the change of bandwidth has on the processed signal output highlights precisely why the narrow-band technique is superior to the overall level technique as a diagnostic tool. For example, a certain transducer may provide an output power spectral density output that may be interpreted in the ways shown in Figure 8.13. It is apparent that the components around frequency f_1 dominate the overall level reading, and the shape of the third octave result. Important changes, say at f_2 and f_3 or the presence of other components, could go largely unnoticed, except by the use of narrow-band methods. This is crucial because the flexibility of the mechanical system may be such that important components are masked by those closer to resonances in the mechanical structure of the machine.

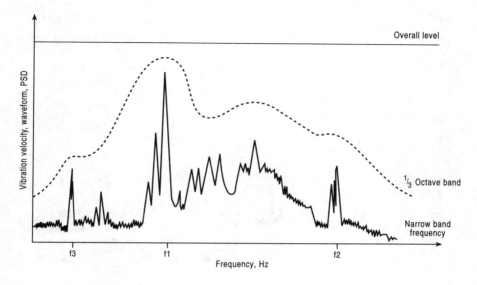

Figure 8.13   Effect of change in bandwidth on spectral response

The narrow band spectrum also allows the operator to trend the condition of the machine most effectively. This requires that an initial baseline spectrum is taken and subsequent spectra are compared with it. The use of digitally derived spectra means that the results of such comparisons can be computed quickly since the spectra reduce to a simple sequence of numbers at discrete frequencies, as closely spaced as required within the limitations of the instrumentation. In this way criteria such as VDI 2056 can be applied for each frequency. Because of the large amounts of data generated using narrow-band methods, it is frequently convenient to predetermine the operational limits, on the basis of one of the vibration standards, and to construct an operational envelope around the baseline spectrum. This can take account of the wider limits allowable at higher frequencies, and can be used to automatically flag warnings when maintenance limits are reached. The basis of this technique is illustrated in Figure 8.14, where the baseline is set at the maximum value of vibration expected and the operational envelope at which trips are initiated, is set above this.

The techniques described thus far in this section are relatively general. In order to identify not just unsatisfactory overall performance, but to pinpoint specific problems, it is necessary to examine discrete frequencies, or groups of frequencies, as indicated earlier in this chapter. Induction motors in particular require a high degree of frequency resolution applied to their vibration signals since the speed of rotation is close to the electrical supply frequency. This tends to generate side bands spaced at s and $2s$ around the harmonics of the supply frequency, where s is the slip frequency of the machine.

The application of vibration monitoring for fault diagnosis in large turbine generators has been described by Mayes [36] and computer analysis techniques that can be applied off-line to vibration data collected on-line are described by Herbert [37].

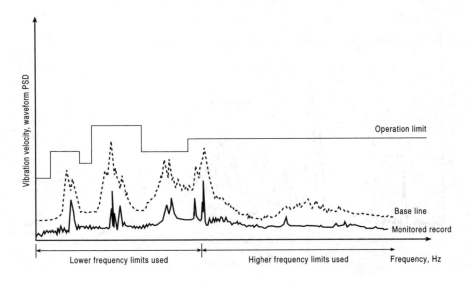

Figure 8.14   Operational envelope around a spectral response

Some of the detail of analysis and the effect of the foundation response to the machine excitation is given by Smart *et al.* [38].

8.6.3   Faults detectable from the stator force wave

Using the techniques described in the previous section and verifying by measurement, it has been shown that UMP can excite stator side vibration components at one, two and four times the supply frequency. Dynamic unbalance and coupling misalignment also produce this effect [11,39].

The last reference also suggests that even orders of the fundamental occur in the frame vibration spectrum due to inter-turn winding faults on the stator [39]. The authors use the principal slot harmonics as an indicator of eccentric running. They also use the transverse mechanical angular frequency, ω_{sm}, measured on the stator frame, excited by radial forces, given by Yang [4] and others as

$$\omega_{sm} = \omega_{se}\left[(nN_r \pm k_e)\frac{(1-s)}{p} \pm q\right] \tag{8.35}$$

where n is an integer, q is the space harmonic of the stator winding MMF, 1, 3, 5, 7 ..., k_e is the so-called eccentricity order number, which is zero for static eccentricity and a low integer value for dynamic eccentricity.

They report tests on a machine with a 51-slot rotor (N_r) and a four-pole, three-phase stator, which exhibits an increase in frame vibration levels of 25 dB and 17 dB at the frequencies 1104 Hz and 1152 Hz due to the introduction of 50 per cent dynamic eccentricity.

There are strong indications that frame vibration can be used to monitor a variety of fault conditions, particularly in induction machines. Caution must be exercised, however, for vibration transmitted from adjacent or coupled plant may excite a natural mode in the machine, while a forced component from a fault within the machine may be sufficiently different from any natural frequency to cause only a slight response.

This effective signal-to-noise ratio for the detection of faults by frame-borne vibration can be quantified using the notion of system flexibility or mobility, the flexibility being regarded as the sum of the modal responses of the machine, and the vibration experienced by the machine is the product of the exciting force wave and this flexibility. Smart *et al.* [38] give an explanation of the interaction of the foundation response of a turbine generator to the exciting effect of unbalance during a rundown.

A number of authors have reported the identification of various vibrational frequencies associated with faults in induction machines, including Rai [11], Dorrell *et al.* [39], Hargis [40] and Nandi *et al.* [41]. It has been suggested by Rai [11] that, on a machine supplied at $f_{se} = 50$ Hz, vibration at or near 50, 100 and 200 Hz is indicative of eccentricity, but the picture is confused because other anomalies also manifest themselves by the production of such frequencies, for example misalignment and dynamic unbalance. Dorrell *et al.* [39] showed that on a machine supplied at $f_{se} = 50$ Hz, the stator frame vibration will exhibit 100, 200 and 300 Hz components due to an inter-turn winding fault or supply voltage unbalance, including single phasing. They also show that higher-order harmonics occur in the stator frame vibration due to eccentricity, as derived from (8.35). However, the exact arrangement of the drive and the nature of the coupled load may be of critical importance, since transmitted vibration may mask the frequency that one is hoping to measure.

Vibration can occur in electrical machinery as a result of both electrical and mechanical action. Finally, Trutt *et al.* [42] recently carried out a theoretical review of the relationships that should exist between electrical winding parameters and the mechanical vibration of AC machine elements under normal and faulted operating conditions. Table 8.5 has been compiled to distil the information to be found in these various references.

8.6.4   Torsional oscillation monitoring

As we have seen previously in Section 8.4.2, there may be a specific need to monitor the torsional behaviour of a long, thin shaft such as on a turbine generator. The direct approach to this problem would be to mount suitable strain gauges on the shaft, together with suitable telemetry, to transmit the gauge output from the rotating reference frame. This has only been done for experimental purposes and is not appropriate for long-term use owing to the harsh operational conditions the transducers would need to withstand. An indirect method of monitoring the torsional responses of shafts have been outlined by Walker *et al.* [26], in which the twist of the shaft system is measured by comparing the angular displacement of the non-drive end of the high pressure turbine shaft with that of the non-drive end of the generator exciter, see Fig. 8.15. The airgap torque produced by the machine is calculated directly from the monitored electrical quantities. The monitor has a modular construction, the two principal elements being concerned with the capture of the mechanical, or torque

Table 8.5 Mechanical frequency components related to specific machine faults

		Rotational angular frequency of the rotor, ω_m, rad/s	Transverse vibration angular frequency on the stator, ω_{sm}, rad/s	Comments
Mechanical faults	Oil whirl and whip in sleeve bearings	—	$\omega_{sm} = (0.43\,to\,0.48)\omega_{rm}$	Equation (8.33). Pressure-fed lubricated bearings onl
	Damage in rolling element bearings	—	$\omega_{sm} = \frac{n_b}{2}\omega_{rm}\left(1 - \frac{d_b}{D_b}\cos\phi\right)$ $\omega_{sm} = \frac{n_b}{2}\omega_{rm}\left(1 + \frac{d_b}{D_b}\cos\phi\right)$ $\omega_{sm} = \frac{D_b}{2d_b}\omega_{rm}\left(1 - \left(\frac{d_b}{D_b}\right)^2\cos^2\phi\right)$ $\omega_{sm} = \frac{1}{2}\omega_{rm}\left(1 - \frac{d_b}{D_b}\cos\phi\right)$	Common source of vibration. Also frequencies in the range 2–60 kHz due to element resonance. Rolling element bearing faults can also be diagnosed by shock pulse method Table 8.1
	Static misalignment of rotor shaft in a synchronous machine	$\omega_{rm} = \omega_{se}/p$	$\omega_{sm} = 2p\omega_{rm}$	Causes static eccentricity
	Unbalanced mass on rotor of a synchronous machine	$\omega_{rm} = \omega_{se}/p$	$\omega_{sm} = \omega_{rm}$	Very common. Unbalance causes dynamic eccentricity of the rotor (see below)
	Dynamic eccentricity in a synchronous machine	$\omega_{rm} = \omega_{se}/p$	$\omega_{sm} = \omega_{rm}$	Dynamic eccentricity causes UMP in an electrical machine

(continued)

	Rotational angular frequency of the rotor, ω_m, rad/s	Transverse vibration angular frequency on the stator, ω_{sm}, rad/s	Comments
Dynamic displacement of shaft in bearing housing in a synchronous machine	$\omega_{rm} = \omega_{se}/p$	$\omega_{sm} = \omega_{rm}, 2\omega_{rm} \ldots$	Causes dynamic eccentricity. Generates a clipped time waveform, due to shaft motion being limited by bearing constraint, therefore produces a high number of harmonics
General expression for static and dynamic eccentricity in an induction machine	$\omega_{rm} = (1-s)n\omega_{se}/p$	$\omega_{sm} = \omega_{rm}[(nN_r \pm k_e)(1-s) \pm pk]$	Equation (8.35). Sidebands at plus or minus slip frequency may occur and components due to UMP (see above)
Commutator faults in DC machine		$\omega_{sm} = 2pk_c\omega_{rm}$ for lap wound $\omega_{sm} = 2k_c\omega_{rm}$ for wave wound	Unbalanced rotor components also generated
Electrical faults	Broken rotor bar in an induction machine	$\omega_{rm} \pm \dfrac{2ns\omega_{se}}{p}$	$\omega_{sm} = \left(\omega_{rm} \pm \dfrac{2ns\omega_{se}}{p}\right)$	Difficult to detect because of small amplitude. Current, speed or leakage field have better detection levels
	Stator winding faults induction and synchronous machines		$\omega_{sm} = p\omega_{rm}, 2p\omega_{rm}, 4p\omega_{rm} \ldots$	Problems can be identified as of electrical origin by removing supply and identifying change. Cannot differentiate winding fault types on vibration alone, current monitoring also necessary

Definitions: ω_{sm} = mechanical vibration frequency on the stator side; ω_{se} = electrical supply frequency; N_r = integer number of rotor slots; N = rotational speed in rev/min; $f\omega_{rm} = \frac{2\pi N}{60}$ mechanical rotational frequency, Hz; $\omega_{rm} = \omega_{se}/p$ for a synchronous machine; $\omega_{rm} = (1-s)\omega_{se}/p$ for an asynchronous machine s = asynchronous machine rotor speed slip, 0–1; p = pole pairs; n = an integer; k_e = eccentricity order, zero for static eccentricity, low integer value 1, 2, 3 … for dynamic eccentricity; k = space harmonic of the stator winding MMF, 1, 3, 5, 7 …; n_b = number of rolling elements; d_b = rolling element diameter; D_b = rolling element pitch; ϕ = rolling element contact angle with races; k_c = number of faulty commutator segments

transient and electrical transient. The torque transient capture unit is triggered by any sudden increase in the airgap torque or by sudden changes in the shaft angular vibration velocities. Similarly, the electrical transient is captured in response to any sudden change in the value of the line currents. The captured data can then be transmitted for further analysis and evaluation. The software receives the captured data and determines the torsional response of the shaft and the associated impact on the fatigue life of the set. The results obtained in this way can be used to plan maintenance intervals on the basis of need, rather than risk catastrophic failure when there has been a high level of system disturbances between fixed outages.

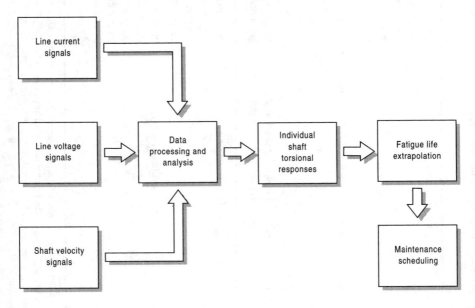

Figure 8.15   Functional description of torsional oscillation monitoring system

The monitoring of torsional oscillations can also be used to detect faults in induction motors. The speed of an induction motor driving an ideal load should be constant. Perturbations in load and faults within the rotor circuit of the machine will cause the speed to fluctuate. If the rotor is defective the speed fluctuation will occur at twice the slip frequency. This is because the normally torque-producing slip frequency currents that flow in the rotor winding are unable to flow through the defective part. In effect the speed fluctuations complement the twice the slip frequency current fluctuation described in Chapter 9. A defective induction machine with a rotor of infinite inertia will have twice the slip frequency current fluctuations and no speed variation, whereas a low inertia rotor will exhibit speed fluctuations but no current fluctuation. Tavner *et al.* reported this method [43] developed by Gaydon., which has been investigated further as the 'instantaneous angular speed' method by Ben Sasi *et al.* [44,45], and a typical measurement result is shown in Figure 8.16.

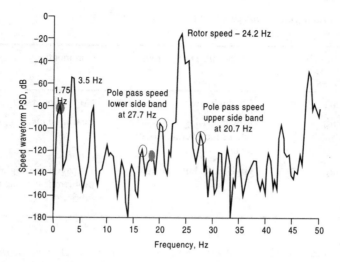

*Figure 8.16   Rotor speed spectrum from an induction motor with a rotor bar fault.
[Taken from Ben Sasi et al. [44]]*

8.6.5   Shock pulse monitoring

The shock pulse method is used exclusively for detecting faults in rolling element bearings and is based on the principles described in Section 8.5.2. Since the largest proportion of electrical machine failures are ascribable to bearing problems (see Table 3.2) and the majority of machines are smaller than 300 kW, this technique is important for a large number of machines. As a rolling element bearing deteriorates the moving surfaces develop small pits or imperfections and the interaction between such surfaces generates mechanical stress waves or shock pulses in the bearing material, which propagate into the frame and structure of the electrical machine. These shock pulses are at ultrasonic frequencies and can be detected by piezoelectric transducers with a strong resonant frequency characteristic tuned to the expected frequency of the pulses at around 32 kHz. To increase the sensitivity of the electronic conditioning, the transducer is tuned to this resonant frequency. A peak hold circuit enables the maximum value of the shock pulse to be recorded, which is taken to be a measure of the condition of the rolling element bearing. This condition of the bearing is assessed by defining a quantity known as the shock pulse value (SPV) defined as:

$$\text{SPV} = \frac{R}{N^2 F^2} \tag{8.36}$$

where R is the shock pulse meter reading, N is the shaft speed in rev/min, and F is the factor relating to bearing geometry. Low values indicate bearings in good condition. Generally the technique is best used in conjunction with overall vibration level monitoring, and Table 8.6 can be used to give qualitative guidance on the bearing condition.

Table 8.6   Shock pulse interpretation

Overall vibration level trend	Shock pulse value trend	Comments
Low and rising	Remains low	No bearing damage
Low and rising	Low but rising at the same rate as the overall vibration level	Bearing damage likely
Low and rising	High value but constant	Damaged bearing but another problem is causing the rising vibration

Tandon *et al.* [46,47] describe the development of rolling element bearing monitoring including vibration, acoustic and shock pulse methods. There is a technique for assessing the thickness of the oil or grease film in the rolling element bearing, based on the experimental evidence that the shock pulse value increases, in approximately the manner shown in Figure 8.17, as a function of the percentage of dry contact time per revolution. The dry contact time was measured by monitoring current flow between the inner and outer races of a test bearing and current flow taken as an indication of dry contact.

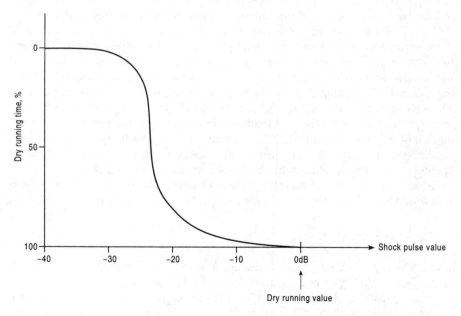

Figure 8.17   Shock pulse value as a function of dry running time per revolution

Stack *et al.* [30] classify motor bearing faults illustrating the unpredictable and broadband nature of the effects produced, emphasising its significance, because a successful bearing condition-monitoring scheme must be able to reliably detect all classes of faults. The quantitative evaluation of bearings using the shock pulse method therefore remains difficult.

8.7 Conclusion

Vibration measurement is at the heart of the monitoring of rotating machines. Although electrical machines are generally low-vibration devices, they may be coupled to high-vibration prime movers or driven plant via flexible couplings and mounted on separate foundations via resilient mounts. The excitation of electrical machine vibration is generally mechanical unbalance or harmonic electromagnetic forces originating from the machine airgap. The response of the machine to these exciting forces depends on the precise coupling and mounting of the machine.

Vibration monitoring and shock pulse analysis are non-invasive but use a number of specialised sensors, broad bandwidth and complex analysis. The precise selection and location of sensors is very important. However, because of its wide application in other rotating machines vibration analysis has established itself as a reliable and widely accepted technique for electrical machines and shock pulse analysis, and also (particularly for bearings) because it is capable of differentiating between mechanical and electromagnetic excitation forces, which is invaluable in detecting root causes before they develop into failure modes. Motor speed has been analysed using instantaneous angular speed to detect rotor electrical faults but has not been widely used by operators. The following chapter will show the close relationship between vibration and electrical monitoring of the machine.

8.8 References

1. Alger P.L. *The Nature of Induction Machines*. New York: Gordon and Beach; 1965.
2. Jordan H. *The Low Noise Electric Motor*. Essen: Springer Verlag; 1950.
3. Erdelyi E. and Erie P.A. Vibration modes of states of induction motors. *ASME Transactions* 1956; **A-28**: 39–45.
4. Yang S. *Low Noise Electric Motors*. Oxford, UK: Clarendon Press; 1981.
5. Timoshenko S. *Vibration Problems in Engineering*. New York: John Wiley & Sons, Van Nostrand Inc.; 1974.
6. Penman J., Chalmers B.J. et al (1981). The performance of solid steel secondary linear induction machines. *IEEE Transactions on Power Apparatus and Systems* 1981; **PAS-100**: 2927–35.
7. Hague B. *Principles of Electromagnetism Applied to Electrical Machines*. New York: Dover; 1962.
8. Stafl M. *Electrodynamics of Electrical Machines*. Prague: Academia; 1967.

9. Binns K.J. and Dye M. Identification of principal factors affecting unbalanced magnetic pull in cage induction motors. *IEE Proceedings, Part B* 1973; **120**: 349–54.

10. Swann S.A. Effect of rotor eccentricity on the magnetic field of a non-salient pole machine. *IEE Proceedings* 1963; **110**: 903–15.

11. Rai R.B. *Airgap Eccentricity in Induction Motors*. ERA Report, 1974: 1174–88.

12. Binns K.J. Cogging torques in induction machines. *IEE Proceedings* 1968; **115**: 1783–90.

13. Lim C.Y. Characteristics of reluctance motors. *IEEE Transactions on Power Apparatus and Systems* 1951; **PAS-70**: 1971–8.

14. Williamson S. and Smith A.C. Field analysis for rotating induction machines and its relationship to the equivalent circuit method. *IEE Proceedings, Part B* 1980; **127**: 83–90.

15. Vas P. *Parameter Estimation, Condition Monitoring and Diagnosis of Electrical Machines*. Oxford, UK: Clarendon Press; 1996.

16. Carpenter C.J. *Surface Integral Methods of Calculating Force on Magnetized Iron Parts*. IEE Monograph No. 342, 1959.

17. Brandl P. Forces on the end windings of AC machines. *Brown Boveri Review* 1980; **2**: 128–34.

18. Ohtaguro M., Yagiuchi K. and Yamaguchi H. Mechanical behaviour of stator endwindings. *IEEE Transactions on Power Apparatus and Systems* 1980; **PAS-99**: 1181–5.

19. Williamson S. and Ellis M.R.E. Influence of rotor currents on end-winding forces in cage motor. *IEE Proceedings, Part B, Electric Power Applications* 1988; **135**: 371–9.

20. Campbell J.J., Clark P.E., McShane I. E. and Wakeley K. Strains on motor endwindings. *IEEE Transactions on Industry Applications* 1984; **IA-20**: 37–44.

21. Wort J.F.G. The fundamentals of industrial balancing machines and their applications. *Bruel and Kjaer Review* 1981; **1**: 3–31.

22. International Organization for Standardization. Mechanical vibration–balance quality requirements for rotors in a constant (rigid) state. Part 1: Specification and verification of balance tolerances. ISO 1940-1. International Organization for Standardization, Geneva, 2003.

23. Smart M.G., Friswell M.I. and Lees A.W. Estimating turbogenerator foundation parameters: model selection and regularization. *Proceedings of the Royal Society A: Mathematical, Physical and Engineering Sciences* 2000; **456**: 1583–607.

24. Dimentberg F.M. *Flexural Vibrations of Rotating Shafts*. London: Butterworths; 1961.

25. Walker D.N., Bowler C.E.J., Jackson R.L. and Hodges D.A. Results of subsynchronous resonance test at Mohave. *IEEE Transactions on Power Apparatus and Systems* 1975; **PAS-94**: 1878–89.

26. Walker D.N., Adams S.L. and Placek R.J. Torsional vibration and fatigue of turbine-generator shafts. *IEEE Transactions on Power Apparatus and Systems* 1981; **PAS-100**: 4373–80.

27. Joyce J. S. and Lambrecht D. Status of evaluating the fatigue of large steam turbine generation caused by electrical disturbances. *IEEE Transactions on Power Apparatus and Systems* 1980; **PAS-99**: 111–19.
28. Cudworth C.J., Smith J.R. and Mykura F. Mechanical damping of torsional vibrations in turbogenerators due to network disturbances. *Proceedings of the Institution of Mechanical Engineers* 1984: 139–145.
29. Collacott R.A. *Vibration Monitoring and Diagnostics.* London: G. Godwin; 1979.
30. Stack J.R., Habetler T.G. and Harley, R.G. Fault classification and fault signature production for rolling element bearings in electric machines. *IEEE Transactions on Industry Applications* 2004; **IA-40**: 735–9.
31. Ehrich F. and Childs D. Self-excited vibration in high-performance turbomachinery. *Mechanical Engineering* 1984; **106**: 66–79.
32. Verein Deutscher Ingenieure. Beurteilungsmaßstäbe für mechanische Schwingungen von Maschinen. VDI 2056. VDI Dusseldorf, Germany: Verein Deutscher Ingenieure 1964.
33. International Organization for Standardization. Mechanical vibration – Evaluation of machine vibration by measurements on non-rotating parts – Part 1: General guidelines. ISO 10816-1:1995. ISO, Geneva, 1995.
34. Canadian Government. Vibration Limits for Maintenance. Specification CDA/MS/NVSH/107. Ohawa, Canada, Canadian Government.
35. Neale N. and Associates. *A Guide to the Condition Monitoring of Machinery.* London: Her Majesty's Stationary Office; 1979.
36. Mayes I.W. Use of neutral networks for online vibration monitoring. *Proceedings of the Institution of Mechanical Engineers Part A, Journal of Power and Energy* 1994; **208**: 267–74.
37. Herbert R.G. Computer techniques applied to the routine analysis of rundown vibration data for condition monitoring of turbine-alternators. *British Journal of Non-Destructive Testing* 1986; **28**: 371–5.
38. Smart M.G., Friswell M.I. and Lees A.W. Estimating turbogenerator foundation parameters: model selection and regularization. *Proceedings of the Royal Society A: Mathematical, Physical and Engineering Sciences* 2000; **456**: 1583–607.
39. Dorrell, D. G., W. T. Thomson and Roach S. Analysis of airgap flux, current, and vibration signals as a function of the combination of static and dynamic airgap eccentricity in 3-phase induction motors. *IEEE Transactions on Industry Applications* 1997; **IA-33**: 24–34.
40. Hargis C. Steady-state analysis of 3-phase cage motors with rotor-bar and end-ring faults. *IEE Proceedings, Part B, Electric Power Applications* 1983; **130**: 225.
41. Nandi S., Bharadwaj R.M., Toliyat H.A. and Parlos A.G. Performance analysis of a three-phase induction motor under mixed eccentricity condition. *IEEE Transactions on Energy Conversion* 2002; **EC-17**: 392–9.
42. Trutt F.C., Sottile J. and Kohler J.L. Detection of AC machine winding deterioration using electrically excited vibrations. *IEEE Transactions on Industry Applications* 2001; **IA-37**: 10–14.

43. Tavner P.J., Gaydon B.G. and Ward D.M. (1986). Monitoring generators and large motors. *IEE Proceedings, Part B, Electric Power Applications* 1986; **133**: 169–80.
44. Ben Sasi A.Y., Gu F., Li. Y. and Ball A.D. A validated model for the prediction of rotor bar failure in squirrel-cage motors using instantaneous angular speed. *Journal of Mechanical Systems and Signal Processing* 2006; **20**: 1572–89.
45. Ben Sasi A.Y., Gu F., Payne, B.S. and Ball A.D. Instantaneous angular speed monitoring of electric motors. *Journal of Quality in Maintenance Engineering* 2004; **10**: 123–35.
46. Tandon N. and Nakra B.C. Comparison of vibration and acoustic measurement techniques for the condition monitoring of rolling element bearings. *Tribology International* 1992; **25**: 205–12.
47. Tandon N., Yadava G.S. and Ramakrishna K.M. A comparison of some condition monitoring techniques for the detection of defect in induction motor ball bearings. *Mechanical Systems and Signal Processing* 2007; **21**: 244–56.

Chapter 9
Electrical techniques: current, flux and power monitoring

9.1 Introduction

Chapter 1 explained that electromechanical protective relays were the earliest electrical technique for monitoring motors. The technology of these devices is dated, their purpose being exclusively to detect gross perturbations in the electrical quantities at the terminals of the machine to protect against catastrophic damage. They are now largely being replaced by digital protection devices detecting voltage and current perturbations, as described in Chapter 1 (see Figure 1.2).

Within the machine the magnetic flux varies, circumferentially in the airgap, periodically in space and, for an AC machine, periodically with time, as described in Chapter 8. Under ideal conditions this magnetic flux waveform will be symmetrical but electrical faults in the machine will distort it. Rotor faults could be detected by electrical sensors fixed to the rotor, and stator faults could be detected by electrical sensors fixed to the stator. Faults on either rotor or stator disrupt the radial and circumferential patterns of flux in the machine causing changes to the power being fed to the machine, which can be detected via its terminal quantities – voltage, current and power – measured outside the machine to give an indication of its condition. In effect, this is a more prolonged and detailed process than that conceived of in a digital protection relay. In the following sections we describe a number of these new techniques.

9.2 Generator and motor stator faults

9.2.1 Generator stator winding fault detection

The most significant technique in this area is on-line discharge detection, which is dealt with in Chapter 10.

9.2.2 Stator current monitoring for stator faults

This work, mostly concerned with motors, is connected with the earlier work described in Chapter 8 on the effect of faults on vibrations, which also considered rotor eccentricity but is now extended to consider stator winding faults, see work by

Rai [1], Dorrell *et al.* [2], Hargis [3] and Nandi *et al.* [4]. The work is also closely associated with the detection of rotor winding faults described in Section 9.4.3. The theoretical work, verified by laboratory experiments was started by Penman *et al.* [5] and continued with Penman and Stavrou [6], who concentrated primarily on stator winding faults. Thomson *et al.* [7–9] then took up the practical application of the work to machines in industrial applications but with the particular intention of detecting rotor eccentricity, which could indicate the deterioration of machine bearings, one of the common failure modes of electrical machines as set out in Chapter 3.

9.2.3   Brushgear fault detection

Brushgear, in those machines that use it, requires a steady maintenance commitment if good performance with the minimum of sparking is to be maintained. Poor performance can be detected by measuring brush or brush-holder temperature but a more direct method would be to detect the radiofrequency (RF) energy generated by sparking, as described by Michiguchi *et al.* [10]. They used a wide-bandwidth dipole antenna connected to an RF amplifier with a bandwidth of 10–100 MHz, the output of which was rectified. The processing electronics measured the area under any pulses of RF power that enter the monitor as a result of sparking activity at the brushes. The monitor thereby produces a chart record showing the average area of sparking pulse and Michiguchi relates this to a 'spark number' indicating an intensity of sparking. Maintenance staff use this indication to decide when brushes should be changed.

9.2.4   Rotor-mounted search coils

We have not found any techniques reported for detecting stator faults by search coils mounted upon the rotor. No doubt the usefulness of this technique is affected by the need to mount expensive instrument slip rings on the rotor and its effectiveness will be limited by the reliability of the measurement brushgear, which, as will be described in the section on shaft voltages, is notoriously poor.

9.3   Generator rotor faults

9.3.1   General

The rotors of large turbine generators are particularly highly rated because of the large mechanical and electrical stresses placed upon them, in particular the high centrifugal forces on the winding and the relatively high temperatures attained in the winding insulation. Consequently that part of the machine is particularly prone to faults that, as stated in Chapter 2, tend to develop over a long period of time. The rotor is also relatively inaccessible both for obtaining signals during running and for removal for repair if a fault is detected. These facts, taken together with the high value of turbine generator plant, have meant that monitoring techniques for generator rotors have been developed to a high degree of sophistication. Some of the techniques described are also applicable to smaller-output machines, but have yet to become fully accepted.

9.3.2   Earth leakage faults on-line

A single earth leakage fault on a generator rotor winding is not serious in itself, because it cannot cause any damage as the earth leakage current is limited to leakage resistance of the excitation supply. However, if two well-separated earth faults occur then large currents can flow, leading to significant damage to the winding, its insulation and the rotor forging. The aim of a rotor earth fault detector is to apply a DC bias voltage to the rotor winding and monitor the current flowing to the rotor body via an alarm relay (see Warrington [11] and Figure 9.1(a)). If such an alarm occurs, many utilities would consider that the machine should be shut down so that the rotor can be investigated. However, operational pressures are such that this is often not possible, and it is

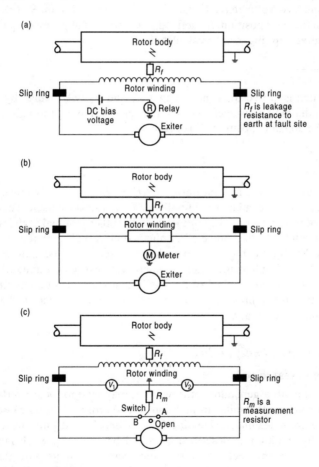

Figure 9.1   *Detecting rotor earth faults. (a) Use of an earth leakage relay. (b) Monitoring of an existing earth fault using a potentiometer. (c) Monitoring for a second earth fault by measuring resistance to earth from each end of the winding.*

necessary to continue running the unit. The next step then is to monitor the earth leakage current and manually trip the unit if there is any further increase, indicative of a second earth fault.

An alternative method is to use a potentiometer fed to earth via a sensitive galvanometer making a bridge circuit, as shown in Figure 9.1(b). As the earth fault location alters or a second fault occurs the bridge unbalances and an indication occurs on the meter. The problem is that the second earth fault may arise close to the location of the first fault and the resultant change in earth leakage current may not be particularly large.

A more sensitive indicator of the onset of a second earth fault is the resistance of the winding to earth, measured from either terminal. Such a technique has been described using two voltmeters, V_1 and V_2, as shown in Figure 9.1(c). When the switch is open the fault position, K, defined as the fractional position up the winding from the negative slip ring, can be calculated:

$$K = \frac{V_1}{V_1 + V_2} \tag{9.1}$$

When the switch is closed to A the voltages V_1 and V_2 will change by an amount depending on the fault resistance, R_f, and the current flowing through the fault, I_f, so that now the apparent position K' is given by:

$$K' = \frac{V_1'}{V_1 + V_2'} \tag{9.2}$$

From the apparent change in fault position $\Delta K = K' - K$, the voltage across the fault resistance can be calculated and finally the fault resistance itself. This procedure can be repeated by connecting the voltmeters to the other terminal of the winding, by closing the switch to B. The choice of terminal connection is governed by the initial fault position, the object being to optimise the measurement of the fault resistance R_f. The scheme can be implemented using a microprocessor-based unit, which makes the measurement at each terminal of the winding at intervals of approximately 1 s, processes the results and presents information for operating staff as well as initiating relay indications if necessary.

9.3.3 Turn-to-turn faults on-line

9.3.3.1 Airgap search coils

Turn-to-turn faults in a generator rotor winding may lead to local overheating and eventually to rotor earth faults. In addition, the shorting of turns causes unequal heating of the rotor leading to bending and an unbalanced pull, which together cause increased vibration as described by Khudabashev [12]. Such faults can be detected off-line by the method of recurrent surge oscillography, described in Section 9.3.4, but a way of detecting them on-line was first described by Albright [13] using a stationary search coil fitted in the airgap of the machine. The search coil, of diameter less than the tooth-width of the rotor, is fixed to the stator usually in the airgap, and detects either the radial or circumferential component of

(a)

(b)

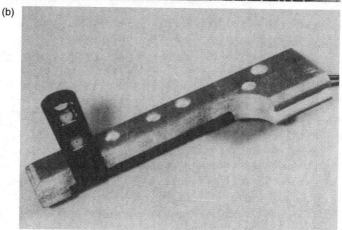

Figure 9.2 Photographs of two typical search coil installations in large generators

magnetic flux. Examples of two types of airgap search coil installation are shown in Figure 9.2.

Figure 9.3 shows typical waveforms obtained from a radial search coil in a two-pole generator operating on load. A normal two-pole rotor will have an even number of winding slots and will produce a radial flux wave, B, in the airgap as follows:

$$B = \sum_{n=1,3,5\,...} B_n \sin n\omega t + \sum_{m\ even} B_m \sin m\omega t \qquad (9.3)$$

This is the normal MMF wave tooth ripple.

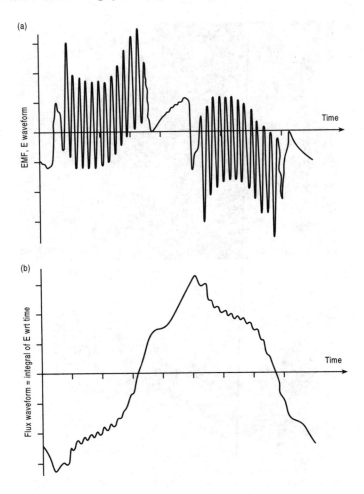

Figure 9.3   Typical voltage and flux waveforms obtained from a generator air gap search coil. (a) Search coil voltage waveform. (b) Flux waveform obtained by integrating (a).

The search coil normal EMF waveform per turn of the search coil will be:

$$e_{normal}(t) = A\frac{dB}{dt}$$

$$= \sum_{n=1,3,5\dots} An\omega B_n \cos n\omega t \text{ normal MMF wave}$$

$$+ \sum_{m\ even} Am\omega B_n \cos m\omega t \text{ normal tooth ripple MMF wave} \qquad (9.4)$$

where A is the effective search coil area. The n odd harmonics are due to the wave shape of the MMF wave in the airgap and are dependent upon the spread of winding slots over the rotor pole pitch. The m even harmonics are due to the rotor tooth ripple that is present in the voltage waveform.

When a shorted turn occurs two things happen. First it disturbs the MMF distribution, causing low-order even harmonics or an asymmetry in the flux and search coil voltage waveforms. Second it disrupts the nth order slot ripple harmonics. This is shown in the search coil faulty waveform as follows:

$$e_{fault}(t) = \sum_{n=1,3,5\ldots} An\omega\, B_n \cos n\omega t \text{ normal MMF wave}$$

$$+ \sum_{\ell=2,4,6\ldots \text{ dependent on fault location}} A\ell\omega\, B_\ell \cos \ell\omega t \text{ fault assymmetric MMF wave}$$

$$+ \sum_{m=even} Am\omega\, B_m \cos m\omega t \text{ normal tooth ripple MMF wave}$$

$$+ \sum_{p=\pm1,\pm3} A(m+p)\omega\, B_n \cos(m+p)\omega t \text{ fault tooth ripple MMF wave}$$

$$(9.5)$$

The heights of the corresponding peaks and troughs in the ripple will change so that the search coil voltage will no longer be symmetrical about zero. In principle, the changes in the heights of the peaks and troughs can be used to determine the number and location of any shorted turns and this is what Albright did in his original paper. He identified faults by measuring the peak heights of the ripple from stored oscilloscope waveforms, recorded under open- and short-circuit test conditions. He did not consider that waveforms obtained with the generator on-load could provide the sensitivity required to detect shorted turns. Since that time a considerable number of large steam turbine-driven generators have been fitted with airgap search coils and a great deal more experience has been obtained of detecting shorted turns. The detection techniques have therefore been refined to deal not only with the different types and locations of search coils but also to detect shorted turns under both off-load and on-load conditions.

New techniques have been developed utilising a digital storage oscilloscope connected to the search coil to give an initial indication of the development of an inter-turn fault. More detailed analysis techniques can then be performed off-line on the stored and downloaded waveforms to positively identify and locate the faults. The on-line method measures the sum of the first four even harmonics of the search coil waveform. The purpose is to identify any asymmetry in the MMF waveform caused by shorted turns. The monitor produces an analogue signal on a chart recorder and the monitor can be adjusted so that any increase above a preset level gives an alarm, which can be used to initiate a more detailed analysis. The setting of that preset level depends upon the generator itself, its history and the type of search coil fitted. Again one is faced with the problems of determining background levels.

The three main methods of detailed analysis of the search coil waveforms yield the following:

- the difference between the search coil voltage waveform and a delayed version of itself;
- the amplitude of the increments in the tooth ripple in the search coil voltage waveform, using the method of Albright [13];
- the flux waveform by integrating the search coil voltage waveform.

Before these can be done the waveform is Fourier analysed into its real and imaginary components. It has been shown that the waveform obtained from a search coil at one radial position in the airgap can be modified to predict the waveform if the coil were at another position closer to the rotor. This is particularly helpful for coils fitted close to the stator surface, where the rotor tooth ripple may be very small. It also allows results from different sizes and designs of machines to be compared on a common basis. The difference waveform can be calculated from the digitised components of the search coil voltage and this waveform can be plotted out to show the presence of a fault as shown in Figure 9.4.

The incremental voltages are calculated by measuring the voltage height between the peaks and troughs of each tooth ripple associated with each pole of the winding and the heights are measured on the side of the ripple furthest from the pole face. These incremental voltages can be plotted out as a histogram over the rotor surface together with a histogram of the differences between the voltages over one pole and the next as shown in Figure 9.5.

The flux waveform is found by integrating the voltage waveform. Distortions of the flux waveform can be brought to light either by direct inspection or by carrying out the difference procedure described above for the search coil waveform. Computer simulations and the practical experience of measuring search coil voltage waveforms suggest that the magnitude of the asymmetry in the search coil waveform produced by a fault depends upon the load as well as the location and number of shorted turns. This is because the degree of saturation affects the magnitude of the rotor tooth ripple, which varies with load and with position around the rotor circumference. In the absence of saturation the asymmetric component would be expected to be proportional to rotor current. However, magnetic saturation of the shorted turn has a significant effect, so that for some loads and locations of shorted turn the magnitude of the asymmetry actually decreases with increasing rotor current.

9.3.3.2   Circulating current measurement

An alternative way of monitoring for sorted turns, which is still under development, uses the stator winding itself as the search coil. The principle of this technique, first suggested by Kryukhin [14], has been developed and fitted to a number of generators in the UK. This technique makes use of the fact that in large two-pole generators, each phase of the stator winding consists of two half-phase windings in parallel. Any asymmetry in the rotor MMF will induce counter-MMF currents in the stator winding with a twice fundamental frequency, which will circulate between the half-phases.

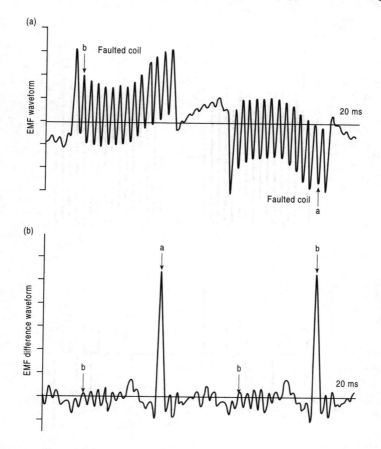

Figure 9.4   *Effect of delaying and adding the waveform from a search coil fitted to a faulty machine. (a) Predicted search coil voltage close to rotor surface. (b) Difference waveform obtained by delaying a) half a cycle and adding to itself.*

The presence of shorted turns is detected by measuring those even harmonic currents. The size of the currents depends upon the severity of the shorted turns, the coupling between rotor and stator and the impedance of the stator winding to the currents. This approach has been developed by others with supporting analysis, for example by Pöyhönen *et al.* [15], but has been subsumed into the more popular analysis of induction motors.

It can be shown that the EMF, $e(t)$, induced across a stator half-phase winding due to a single shorted turn, spanning an angle, 2β, on a two-pole rotor is given by:

$$e(t) = \frac{4\mu_0\omega}{\pi Cg} NI \; r_{airgap} L \sum_{n=1,3,5\ldots}^{\infty} \frac{1}{n} k_{wn} \sin n\beta \cos(n\omega t) \tag{9.6}$$

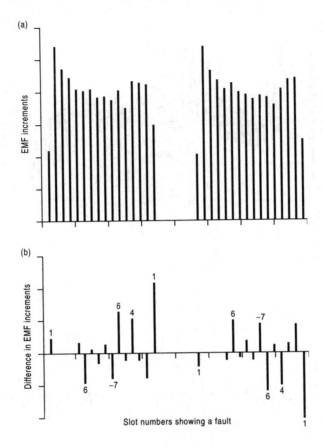

Figure 9.5 *Method of incremental voltages proposed by Albright [13]. (a) Incremental voltages obtained from previous figure. (b) Differences in incremental voltages between poles obtained from Figure (a) above.*

where NI is the ampere-turns of the short, ω is the rotational frequency, N is the number of stator turns in series per half phase, r_{airgap} is the mean radius of the airgap, L is the active length of the rotor, C is the Carter factor to account for slotting, g is the width of the airgap, β is the half angle subtended by the shorted turn, n is the harmonic number, and k_{wn} is the stator winding factor for the nth harmonic.

The EMF induced in the opposite half-phase winding, on the same phase, will be of the same form but the term $\cos(n\omega t)$ will be replaced by one of the form $\cos(n\omega t - n\pi)$. The odd harmonics of the EMFs in the two half-phases will therefore be of the same sign and so when the half-phases are connected in parallel they will aid one another, forming the terminal voltage due to that shorted turn. The even harmonics will be of the opposite sign and so will drive circulating currents between the half phases. The currents are flowing in the stator winding, rotating at the same speed as the rotor, effectively replacing the rotor shorted turn. The second harmonic circulating current that flows in the stator, i_{2C}, can be related to the second harmonic EMF, e_2, induced by the shorted turn, providing the second harmonic impedance, X_2, of the winding is known. The impedance to second harmonic currents, X_2, is given by

$$X_2 = X_{m2} + X_{\ell2} \tag{9.7}$$

where X_{m2} is the second harmonic magnetising reactance and $X_{\ell2}$ is the second harmonic leakage reactance.

For a typical large machine it has been shown that $X_2 = 0.516$ pu. Therefore for a single shorted turn spanning an angle of 2β on the rotor the second harmonic current circulating in the stator winding is approximated by

$$i_{2C} = \frac{e_2 \sin 2\beta}{0.516} \tag{9.8}$$

The currents are detected using air-cored Rogowski coils wrapped around the winding and a diagram of an on-line monitor for doing this is shown in Figure 9.6.

An advantage of this new technique, when compared to airgap search coils, is that the current transducers can be installed without the need to remove the rotor from the generator. For many generators the half-phase windings are joined within the cooling pressure casing in a fairly restricted space that requires special arrangements to gain access. However, on some machines the half-phase windings are joined outside the casing, so fitting of the Rogowski coils becomes simpler. Care must of course be taken to provide appropriate high-voltage insulation and electrostatic screening between the Rogowski coil and the conductor. A disadvantage of the circulating current method, however, is that it does not give information on the turn location, whereas the airgap search coil method does.

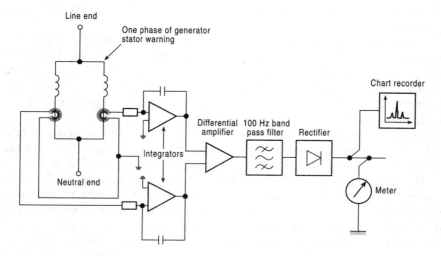

Figure 9.6   Continuous monitor for use on circulating current Rogowski coils

Neither the airgap search coil nor the Rogowski coil methods appear to have been applied to multi-pole hydro-type generators or even four-pole turbine-type machines. No doubt applications will evolve as operational circumstances demand them.

9.3.4   Turn-to-turn and earth leakage faults off-line

Surge techniques have been used for many years by transformer manufacturers for locating faults in windings. More recently turbine generator manufacturers have used similar techniques for pinpointing faults in their rotor windings, as a quality control check immediately after a rotor has been assembled. Work in the former Central Electricity Generating Board showed that such a technique can be used to detect both earth leakage and turn-to-turn faults on generators during their service lives, which used a mercury-wetted contact relay to develop recurrent, rapid rise time (circa 20 ns) surges that were injected into the winding between the slip-ring and the earthed body of the rotor. Figure 9.7(a) shows this recurrent surge generator, which is switched on and off at a frequency of 50 Hz, while Figure 9.7(b) shows the connection of the generator to the rotor winding. The winding approximates to a simple transmission line where the propagation is dominated by the geometry and insulation of the winding conductor in the rotor slot. Mutual coupling between turns of the winding will cause dispersion but this effect has been found to be small in solid-steel rotors. When the surge is injected at one end it has a magnitude determined by the source impedance, R, of the recurrent surge generator and the surge impedance of the winding, Z_0. The surge propagates to the far end of the winding in a time, t, determined by the length and propagation velocity of the winding. The surge is reflected at the far end, its magnitude determined by the reflection coefficient, k_r. For a winding with the far end open-circuited, $k_r = 1$ and with it short-circuited $k_r = -1$. The reflected surge returns to the source and if the source impedance equals the surge impedance of the winding ($k_r = 0$ at source), it is absorbed without further reflection. This is shown in Figure 9.8(a).

When an insulation fault to earth or a turn-to-turn fault is encountered, this reflection pattern will be disrupted and may be observed on the oscilloscope. The pattern on the oscilloscope is known as a recurrent surge oscillograph (RSO) and this has become the name of the technique. The rise time of the surge will affect the method's sensitivity and must be less than the propagation time of the surge-front through a single winding turn for sharp reflection to occur.

Figure 9.7(b) shows results obtained from the practical application of this method. The source impedance R was adjusted to be equal to the winding surge impedance and a surge of between 10 and 100 V was applied with a rise time of 20 ns.

The surge was injected into each end of the winding, with the far-end open-circuited, and two reflection traces were obtained as shown in parts (i) and (ii) of Figure 9.8(b), either of which should be compared to the ideal response of part (i) of Figure 9.8(a). The results show the distortion of the trace as a result of dispersion and the lack of surge sharpness. Faults are indicated by superimposing the two traces and observing any deviations between them, as shown in parts (ii) and (iii) of Figure 9.8(b).

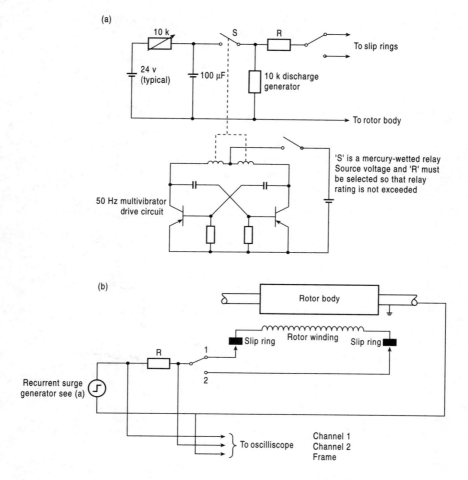

Figure 9.7   *Arrangements for obtaining a recurrent surge oscillograph from a rotor winding. (a) Recurrent surge generator. (b) Connection of the recurrent surge generator to a rotor winding.*

Earth faults may be approximately located by taking the ratio of the times for reflections from the fault and from the end of the winding. Similarly the electrical length of the winding can be estimated by measuring the time for the surge to make a single pass through the winding. This can be done by short-circuiting the far end of the winding to obtain a trace like that in part (ii) of Figure 9.8(a). Any shortening of the length indicates that shorted turns are present in the winding.

The surge impedance of a rotor winding lies between 20 and 30 Ω. A deviation becomes observable on the RSO trace when the impedance to earth of the fault is a significantly small multiple of the surge impedance. In general, the technique can detect faults with impedances to earth of less than 500–600 Ω. Similarly the technique

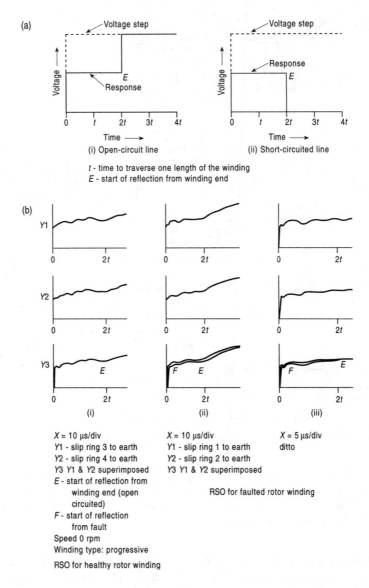

Figure 9.8   Ideal and typical practical recurrent surge oscillographs obtained from generator rotor windings. (a) Applied voltages to rotor winding. (b) Typical responses for rotor winding.

can detect a turn-to-turn fault that has a resistance significantly less than the surge impedance, say of the order of 10 Ω down to zero.

To carry out RSO measurements the generator must be isolated, the field de-energised and the exciter connection disconnected. It is known that many rotor faults are affected by both gravitational and centrifugal forces upon the conductors during rotation, so it is common to carry out the tests stationary, at barring and at speed for comparison purposes. The test has now been applied by many utilities as a routine technique for assessing the operational state of machine windings, although it is clear it cannot be used for brushless machines without the rotor being stationary.

The advantages of the technique are that with the oscillographs it provides a permanent record of the state of a rotor throughout its service life and it can detect both earth and inter-turn faults before their resistance falls to a value where large fault currents flow. Experience has shown that the deterioration of a winding can be assessed by a comparison between oscillographs obtained from the same rotor. It is not possible, however, to make comparisons between oscillographs from different rotors, even if they are of the same design, because of the effects on the surge propagation of quite small variations between the insulation and rotor body properties of different rotors.

The disadvantages are that it detects faults of high resistance and is unable to differentiate between faults that are operationally significant and those that are not. Also, it cannot be used on-line, testing the winding under truly operational mechanical and thermal conditions. This is in contrast to the effectiveness of the airgap search coil or circulating current method that can provide on-line information. There are a number of modern references that cover this and other techniques including Ramirez-Nino and Pascacio [16] and Streifel *et al.* [17].

9.4   Motor rotor faults

9.4.1   General

The rotors of other electrical machines can be highly stressed, though perhaps not to the same degree as turbine generators. Table 2.4 from Chapter 2 shows that large induction motors can have mechanical, electrical or magnetic loadings higher than larger machines and squirrel cage or wound rotors have had problems, as the experience of Tavner *et al.* [18] shows. A number of both electrical and mechanical techniques have developed to monitor these problems. This section deals with the electrical techniques.

9.4.2   Airgap search coils

The work on airgap search coils, described in Section 9.3.2, was all on turbine generators but there is no reason why such methods could not be successfully applied to smaller machines. Indeed Kamerbeek [19] has successfully used the method experimentally on small induction motors but for measuring torque rather than machine faults. A paper by Seinsch [20] has applied this technique to induction motors using a distributed coil on the stator.

9.4.3   Stator current monitoring for rotor faults

Although the technique of using a stator search coil has not been widely used, it is possible to use the stator winding itself as a search coil, in a somewhat similar way to the method described for generators in Section 9.3.3. Any rotor fault in an induction motor will cause a characteristic swing in the supply ammeter reading, which maintenance staff have come to recognise as indicating that trouble is on its way. Careful measurement of the stator current will therefore enable such a fault to be monitored.

The current drawn by an ideal motor should have a single component of supply frequency. Changes in load will modulate the amplitude of the current to produce side bands. Faults in the rotor circuit will generate a side band below the supply frequency and displaced from it by twice the slip frequency. This effect was described in the references in Tavner *et al.* [18] and an explanation is given here. A motor winding with p pole pairs and supply frequency ω_{se} produces a fundamental stator radial MMF wave, f_1, at mechanical angle θ_1 containing odd harmonics only. Consider the fundamental MMF wave

$$f_1(t) = N_1 I_1 \sin(\omega_{se}t - p\theta_1) \tag{9.9}$$

where N_1 is the number of stator turns and I_1 is the stator current.

The angle, θ_2, on the rotor is given by

$$\theta_2 = \theta_1 - \omega_{rm}t \tag{9.10}$$

where ω_{rm} is the angular speed of the rotor and the rotational speed $N = 60\omega_{rm}/2\pi$, so that for a p pole pair rotor the rotor sees the MMF:

$$f_1(t) = N_1 I_1 \sin[(\omega_{se} - p\omega_{rm})t - p\theta_2] \tag{9.11}$$

This MMF rotates forward with respect to the rotor at the slip speed; however, under normal circumstances the rotor carries induced currents, which establish a fundamental rotor MMF wave, f_2, to counter the stator MMF and move at the same speed:

$$f_2(t) = N_2 I_2 \sin[(\omega_{se} - p\omega_{rm})t - p\theta_2] \tag{9.12}$$

If the rotor has a fault, such as a broken bar, the MMF due to the rotor current is modulated by $\sin 2p\theta_2$ so that:

$$f_2(t) = N_2 I_2 \sin[(\omega_{se} - p\omega_{rm})t - p\theta_2]\sin 2p\theta_2 \tag{9.13}$$

Therefore for a p pole pair machine:

$$f_2(t) = \frac{N_2 I_2}{2}\{\cos[(\omega_{se} - pw_{rm})t - 3p\theta_2] - \cos[(\omega_{se} - pw_{rm})t + p\theta_2]\} \tag{9.14}$$

Referring this MMF to the stator, as the counter to (9.11), using (9.10) gives

$$f_2(t) = \frac{N_2 I_2}{2}\{\cos[(\omega_{se} + 2p\omega_{rm})t - 3p\theta_1] - \cos[(\omega_{se} - 2p\omega_{rm})t + p\theta_1]\}$$

$$\tag{9.15}$$

which if we use the fractional slip

$$s = \left(\frac{\omega_{se} - p\omega_{rm}}{\omega_{se}} \right)$$

for a p pole pair induction machine gives

$$\theta_2 = \theta_1 - \frac{(1-s)}{p}\omega_{se}t \tag{9.16}$$

$$f_1(t) = \frac{N_2 I_2}{2}\{(\cos(3 - 2s)\omega_{se}t - 3p\theta_1) - (\cos(1 - 2s)\omega_{se}t - p\theta_1)\} \tag{9.17}$$

Note that these fundamental MMF wave equations (9.11) and (9.17) echo equations (8.22) and (8.23) in Chapter 8 where all harmonic MMFs were considered in the excitation of stator core vibrations by the airgap MMF wave acting on the bore of the core.

The first component of MMF in (9.17) induces zero-sequence EMFs in the three-phase stator winding, because it contains $3\omega_{se}t$ and $3\theta_1$, and gives rise to no current contribution from the supply. The second component of MMF, however, induces a proper three-phase set of currents at the normal supply frequency but contains a component, or side band, $2s\omega_{se}$ below that frequency.

This is twice the slip frequency modulation of the supply current that is seen as the swing on the ammeter reading. Such a cyclic variation in the current reacts back onto the rotor to produce a torque variation at twice the slip frequency that, if the rotor does not have an infinitely high inertia, gives rise to the $2sp\omega_{rm}$ variation in speed or $2s\omega_{se}$ variation in mechanical vibration, that can also be used for fault detection as described in Chapter 8, Section 8.6.4. This speed effect reduces the lower side band, $(1 - 2s)\omega_{se}$, current swing and produces an upper side band at $(1 + 2s)\omega_{se}$, enhanced by modulation of the third harmonic flux in the stator and it can be shown that other side bands at $(1 \pm 2ns)\omega_{se}$ are also found. The ratio of the lower side band amplitude to the main supply frequency component gives a straightforward indication of the extent of rotor damage, as described by Jufer and Abdulaziz. [21].

The supply current can be monitored very easily, without interfering with the machine, simply by fitting a clip-on current transformer around the supply cable to the motor or around the cable of the protection current transformer used to monitor the motor current, see Figure 9.9. The normal procedure is to use a spectrum analyser package in a PC connected via an A/D converter to the current transformer. Surveys of the supply currents to a number of motors can be taken at regular intervals, or when a fault is suspected. Figure 9.10 shows the power spectral density (PSD) for the current from two identical machines. The motor in Figure 9.10(a) had a rotor fault corresponding to three fractured cage bars, shown by spectral components at about 48, 49, 51 and 52 Hz; that is, a slip frequency of 0.5 Hz for $n = 1$ or 1 Hz for $n = 2$ and a slip s of 1 per cent, with side bands described by $(1 \pm 2ns)f_{se}$. The lower side band, due to the MMF modulation, can clearly be distinguished from the supply frequency and an estimate of the fault severity can be made by taking the ratio between the amplitudes of the lower side band and the fundamental frequency.

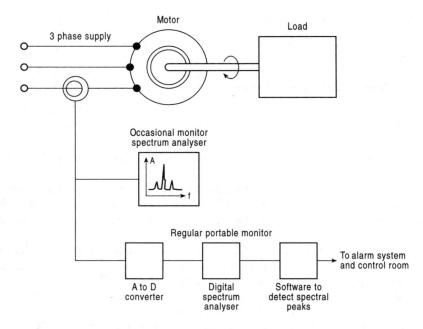

Figure 9.9   Detecting side bands in the supply current of an induction motor

Because the current measuring technique looks into the motor from the terminals it is also possible to see beyond the electrical circuits and detect faults on the mechanical load train such as worn gear teeth, which the motor is driving. Figure 9.10(a) shows a wider frequency range current spectrum from the motor, with no $(1 - 2s)$ component, indicating rotor damage, but other side bands due to the motor driving a load with a damaged gearbox.

Interpretation of the current spectrum requires a relatively skilled operator to carry out a survey of machines and it cannot be considered as continuous monitoring. Normal practice would be to carry out a survey whenever ammeter swings indicate that a problem is imminent. Where motors of high value are at risk, more frequent or continuous monitoring may be necessary. Detection can be more difficult when the motor speed is varying rhythmically because of the driven load, such as a belt or mill drive, or if the frequency variation is significant, for example on a relatively small power system.

The technique has stimulated a surge in investigations in the literature as analysts seek to describe the precise conditions under which faults can be detected. Examples include Menacer *et al.* [22] and Li and Mechefske [23], who make a comparison between current, vibration and acoustic methods of detection.

9.4.4   Rotor current monitoring

The rotor circuits of wound rotor motors are usually poorly protected in most installations. Faults in brazed joints and slip-ring connections have sometimes caused severe

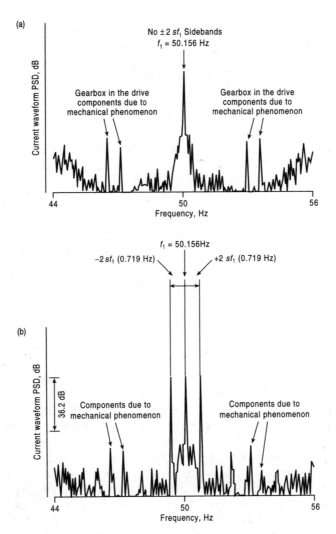

Figure 9.10   *Supply current spectra from induction motor drive trains: (a) with a fault in an attached gearbox; (b) with a rotor cage defect. [Taken from Thomson et al. [9]]*

damage because they have not been detected promptly. Overheating of rotors can also be caused by current imbalance in the external resistors or circuits connected to the slip rings. The low frequency of these currents makes measurements with conventional current transformers inaccurate. Faults of these types were some of the reasons that encouraged the development of proprietary leakage flux technique that is described later in Section 9.5 [24]. However, low-frequency currents can be measured accurately by Rogowski coils. These have been used to monitor the rotor resistance

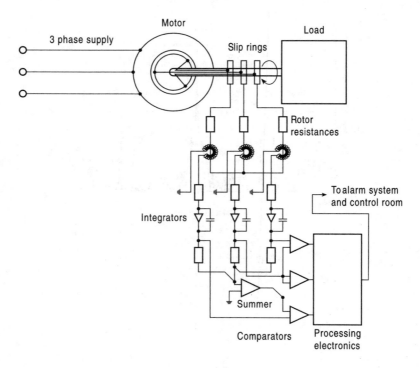

Figure 9.11   Continuous monitor of rotor current in a wound rotor induction motor

currents in variable speed wound rotor motors as shown in Figure 9.11. The signals from the Rogowski coils are integrated to give a voltage proportional to rotor current. These signals are summed to give the mean current in all three phases of the rotor and are compared to the individual phase currents. Processing electronics then detects whether severe unbalance is present and provides amplitude and alarm signals to the control room. The approach described in Reference 24 is based upon a protection philosophy, in that indications from the monitor are used to trip the machine. In practice, however, the instrument has been used to provide monitoring indications that assist in determining when motors should be taken out of service for repair.

9.5   Generator and motor comprehensive methods

9.5.1   General

The reader should be able to see, both from the introduction to this chapter and from the methods that have been described, that electrical techniques have much in common. There would seem to be some advantage in devising a single electrical technique that is capable of detecting all electrical faults, whether they are on the rotor or stator. Trutt *et al.* [25], Kohler *et al.* [26] and Sottile *et al.* [27] have advocated this generalised approach to the monitoring of the terminal quantities of induction machines and Sottile *et al.* [28] have applied the same technique to three-phase synchronous generators. We describe four techniques that detect faults by measuring the effect they have on the machine terminal quantities.

9.5.2 Shaft flux

Shaft flux, or more generally axial leakage flux, occurs in all electrical machines. It is produced because no machine can be constructed with perfect symmetry. There will always be, for example, slight differences in the reluctance of magnetic circuits due to building tolerances, core-plate anisotropy, and plate thickness variation [29]. This asymmetry is reflected in the impedances presented by the various phase groups, or coils in the machine stator, and will cause slight variations between the currents flowing in the coils. It is also the cause of homopolar fluxes in the machine shaft that can lead to shaft voltages as described in a later section.

This asymmetry, together with small differences in the electrical properties of the conductors and variations in the physical disposition of the conduction in both the active length and end regions of machines, will give rise to a net difference between the currents flowing in one section of the end winding when compared with the corresponding section diametrically opposite. The imbalance leads naturally to a net axial flux component. A similar argument can be applied to the rotor circuits; hence one can expect to measure axial flux, even in machines that are in 'perfect health'.

It is a simple extension of this to consider what happens when certain fault conditions arise in a machine. Faults, such as winding short circuits, voltage imbalance and broken rotor bars, represent severe disruptions to the internal symmetry of the machine. It is logical to conclude, therefore, that the effect on the production of axial flux will be readily observable. Any gross change of magnetic circuit conditions, such as the formation of an eccentric airgap due to bearing wear, will, by the same token, be reflected with a corresponding change in axial leakage flux.

The purpose of axial flux monitoring is therefore to translate observed differences in the nature of the axial leakage flux into an indication of fault condition. The production of such fluxes in squirrel cage rotor induction machines was studied by Jordan *et al.* [30,31] with particular emphasis on the changes occurring due to static eccentricity. Erlicki *et al.* [32] showed that it is possible to detect the loss of a supply phase through axial flux monitoring.

In the 1970s in the UK a large power station boiler auxiliary induction motor, equipped with two stator windings for two-speed operation, was operating on one winding with a shorted turn on the idle winding. Circulating currents in the faulty idle winding, induced by the energised winding, caused degradation of the insulation, charring and the generation of flammable gases (see Chapter 7, Section 7.2). Eventually these gases entered the terminal box, ignited, bursting the terminal box and causing a fatality. This incident galvanised interest in the UK in motor monitoring, because it was quickly realised that the faulted idle winding could have been detected by monitoring the axial leakage flux of the machine, particularly by mounting a search coil around the rotor shaft.

Rickson [24] developed a protection device based on this principal. Penman *et al.* [33] showed that more discrimination could be achieved between a variety of fault conditions by carefully processing the axial flux signal and this initiated Penman's work on machine condition monitoring.

The technique relies upon examining the changes in the spectral components of the axial flux. These components arise as described below. Since the fluxes are produced by winding currents, the frequency of these flux components must be related to the frequencies of the currents. Rotor currents are also induced by the airgap flux, so the net airgap flux will be modified as a result. While the rotor is at rest the airgap field results solely from the currents flowing in the stator; hence only the time harmonics present in the line currents will appear in the axial flux. Once the rotor moves, however, it does so with an angular speed, $\omega_{rm} = (1 - s)\omega_{se}/p$, with respect to the stator, where p is the number of pole pairs in the machine. The airgap flux components will consequently be frequency filtered. For example, in the normal three-phase stator winding, the airgap field produced, b_{stator}, can be approximated up to the seventh harmonic by the form

$$b_1(t) = \hat{B}_1 \cos(\omega_{se}t - p\theta_1) + \hat{B}_5 \cos(\omega_{se}t + 5p\theta_1)$$

$$- \hat{B}_7 \cos(\omega_{se}t - 7p\theta_1) + \cdots \tag{9.18}$$

We can transform this expression into a frame of reference moving with the rotor by considering Figure 9.12, which shows the relationships between a fixed point in the stator and a fixed point on the rotor:

$$\theta_2 = \theta_1 - \omega_{rm}t \tag{9.10}$$

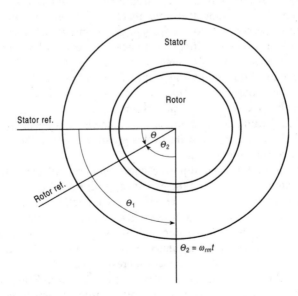

Figure 9.12   Rotor and stator frames of reference

However, for an induction machine with p pole pairs:

$$\phi_2 = \theta_1 - \frac{(1-s)}{p}\omega_{se}t \qquad (9.16)$$

Using these expressions it can be shown that the nth term of the airgap field in the stator frame is

$$b_{n1}(t) = \hat{B}_n \cos[(1 \pm (1-s)n\omega_{se}t) \pm np\theta_1] \qquad (9.19)$$

The rotor frame expression corresponding to (9.18) is

$$b_2(t) = \hat{B}\cos(s\omega_{se}t - p\theta_2) + \hat{B}_5\cos((6-5s)\omega_{se}t - 5p\theta_2)$$
$$- \hat{B}_7\cos((7s-6)\omega_{se}t - p\theta_2) + \cdots \qquad (9.20)$$

The first airgap harmonic produces currents at s times the supply frequency; the fifth airgap harmonic produces time frequencies of $(6-5\ s)$ times and so on.

It is now apparent that the axial flux spectrum is rich in harmonics, even in a well-constructed, healthy machine. Moreover, because fault conditions such as shorted turns, loss of phase, eccentricity and so on, cause changes in the space harmonic distributions in the airgap, such conditions will be accompanied by a corresponding change in the time harmonic spectrum of axial flux. Furthermore, by effectively using the stator winding as a search coil to detect rotor faults, and the rotor winding to detect stator faults, it is possible to gain insight into the harmonic changes to be expected for a given fault condition.

Let us follow a typical fault condition through the diagnostic procedure. If we assume that an inter-turn short circuit exists in the stator winding then this condition can be represented as a single pulse of MMF, similar to that shown in Figure 9.13.

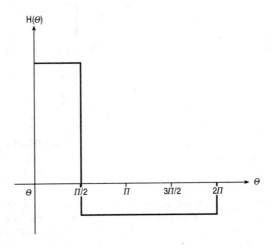

Figure 9.13   MMF due to a single fully pitched coil

The components of the stator airgap field b_{stator} generated by this distribution are

$$b_1(t) = \sum_{n=1,3,5\ldots} \hat{B}_n \cos(\omega_{se}t \pm n\theta) \tag{9.21}$$

The nth component in the rotor frame will therefore be

$$b_{n,2}(t) = \hat{B}_n \cos \left[\left(1 \pm (1-s)\frac{n}{p} \right) \omega_{se}t \pm n\theta_2 \right] \tag{9.22}$$

These harmonics will induce currents in the rotor circuits, and because there are asymmetries in the rotor magnetic and electric circuits, they will appear as additional components in the spectrum of axial flux. Table 9.1 at the end of the chapter summarises the angular frequency components arising in the axial flux. Figure 9.14

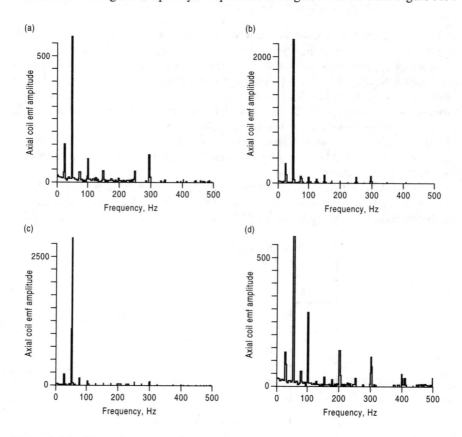

Figure 9.14   *Typical spectra taken at identical gains from an axial flux search coil fitted to an experimental motor. (a) Good rotor, no faults, no load; (b) Broken rotor bar, no other faults; (c) Good rotor, large stator shorted turn 1 amp, no load; (d) Good rotor, small stator negative phase sequence, no load.*

illustrates comparable results from a small four-pole squirrel cage induction machine using the technique. Only the spectral components below 500 Hz are shown, but faults, such as inter-turn short circuits, broken rotor bars, or negative phase sequence in the supply, are visible in the spectra and have been identified.

The axial flux monitoring technique is still embryonic but essentially it requires the collection of an axial flux signal, using a search coil wound concentrically with the shaft of a machine. This signal is then spectrally analysed and on the basis of the appearance of certain harmonic groups a decision is made as to the condition of the machine. The attractions of the method are that it is completely non-invasive and a single sensor can be used for a variety of fault types. It is, however, a complicated technique requiring specialised equipment, and is relatively untested.

9.5.3   *Stator current*

Stator current has been shown in Sections 9.2.2 and 9.4.3 to be a viable condition-monitoring technique for detecting faults in electrical machines. The results of Figure 9.10(a) have also shown that it also has the capability to look beyond the electrical machine itself and detect faults in the mechanically driven load. A further example of fault detection in the driven machine by current analysis was shown, for a variable speed drive downhole pump, in Figure 5.10 by Yacamini *et al.* [34] and Ran *et al.* [35], where wavelet analysis was used to deal with the non-stationary behaviour of the signal.

There is now a more extensive literature on the current analysis method including Joksimovic and Penman. [36] and Stavrou *et al.* [37] who investigated the fundamental magnetic field effects on the airgap, Bellini *et al.* [38,39] who give a good account of the present state of the art and Heneo *et al.* [40,41]. Table 9.2 summarises the angular frequency components that can be detected in the stator current and their relation to machine faults.

9.5.4   *Power*

Recent work by Trzynadlowski *et al.* [42] has shown that the power spectrum may be an effective monitor of machine health and it may simplify some of the complexities of the stator current and axial flux spectra. This was a technique pioneered by Hunter *et al.* [43] and the process can be considered as follows. The instantaneous power, p_1, delivered to or from a three phase machine is given at the stator terminals by

$$p_1(t) = \sum_{q=1}^{3} \sum_{n=1}^{\infty} \sum_{m=1}^{\infty} \hat{V}_{nq} \sin(\theta_{nn}) \hat{I}_{mq} \sin(\theta_{mm} + \phi_{mq}) \tag{9.23}$$

where θ_{nn} or θ_{mm} are defined as follows:

$$\theta_{\ell k} = \ell \omega_{se} t + (q - 1) \frac{2\pi k}{3} \tag{9.24}$$

where $\hat{V}_{nq}$ is the peak of the nth harmonic of the phase voltage in the qth phase assuming that the phase voltages are angularly spaced by $2\pi/3$, $\hat{I}_{mq}$ is the peak of the

m^{th} harmonic of the phase current in the qth phase assuming that the phase currents are angularly spaced by $2\pi/3$, q is the phase number, n is the voltage harmonic number, m is the current harmonic number, and ϕ_{nq} is the phase angle between the phase current, $\hat{I}_{nq}$, and the phase voltage, $\hat{V}_{nq}$.

If the load on the machine is perfectly balanced between the three phases and the voltages and currents contain no harmonics then the expression for instantaneous power reduces to

$$p_1(t) = \hat{V}_1 \hat{I}_1 \sum_{q=1}^{3} \sin(\theta_{11} + \phi) \tag{9.25}$$

$$p_1(t) = \frac{\hat{V}_1 \hat{I}_1}{2} \sum_{q=1}^{3} \{\cos\phi - \cos(\theta_{2,2} + \phi)\} \tag{9.26}$$

The second term in brackets sums to zero because the three phase components are spaced in phase at $4\pi/3$ radians. The net balanced power delivered to or from the stator of the machine is therefore as expected:

$$P_1 = \frac{3}{2}\hat{V}\hat{I}\cos\phi \tag{9.27}$$

Consider the case where the peak phase voltages are balanced and include harmonics, V_n, but the phase currents are unbalanced and also include harmonics. Equation (9.25) can be rewritten in terms of the peak positive, negative and zero sequence harmonic currents, $\hat{I}_{+m}$, $\hat{I}_{-m}$ and $\hat{I}_{0m}$ respectively, as follows:

$$p_1(t) = \sum_{n=1}^{\infty}\sum_{m=1}^{\infty} \hat{V}_n \sin(\theta_n)\{\hat{I}_{+m}\sin(\theta_{+m} + \phi_{+m})$$
$$+ \hat{I}_{-m}\sin(\theta_{-m} + \phi_{-m}) + \hat{I}_{0m}\sin(\theta_{0m} + \phi_{0m})\} \tag{9.28}$$

where in this case:

$$\theta_n = n\omega_{se}t$$
$$\theta_{+m} = +m\omega_{se}t$$

Equation (9.28) reduces to

$$p_1(t) = \sum_{n=1}^{\infty}\sum_{m=1}^{\infty} \left\{ \begin{array}{l} \dfrac{\hat{V}_n\hat{I}_{+m}}{2}\left[\cos(\theta_{(m-n)(m-n)} + \phi_{+m}) - \cos(\theta_{(m+n)(m+n)} + \phi_{+m})\right] \\[2ex] +\dfrac{\hat{V}_n\hat{I}_{-m}}{2}\left[\cos(\theta_{-(m-n)(m-n)} + \phi_{-m}) \right. \\[1ex] \left. \qquad\qquad - \cos(\theta_{-(m+n)(m+n)} + \phi_{-m})\right] \\[2ex] +\dfrac{\hat{V}_n\hat{I}_{0m}}{2}\left[\cos(\theta_{(m-n)-n} + \phi_{0m}) - \cos(\theta_{(m+n)+n} + \phi_{0m})\right] \end{array} \right\} \tag{9.29}$$

When $m = n$ a DC contribution is made to the power in each phase as shown in (9.27). For harmonic components of power, however, because of the presence of the phase factor, $(q - 1)2\pi/3$, a contribution to the power only occurs when $(m - n)$ or $(m + n)$ are multiples of 3, or are triplens. These contributions will be at $\pm(m - n)\omega_{se}t$ or $\pm(m + n)\omega_{se}t$, depending on whether the positive or negative sequence currents are contributing to the power.

From inspection of the last term in (9.29) it can be seen that there will be no zero sequence contribution to the ripple in instantaneous power. This is because for all values of n the term in $(q - 1)2\pi/3$ in θ ensures that the summation over three phases always comes to zero.

An application of these equations to the power condition monitoring of an electrical machine would be a three-phase induction motor with a broken cage. We already know from Table 9.1 that at a fundamental supply voltage angular frequency of ω_{se} the fundamental supply current will contain components at $(1 \pm 2s)\omega_{se}$.

Therefore for the first and $(1 - 2s)^{\text{th}}$ harmonics of voltage and current respectively

$$\pm(m - n)\omega_{se} = \pm2s\omega_{se}$$
$$\pm(m + n)\omega_{se} = \pm2(1 - s)\omega_{se}$$

So the power spectrum due to the fundamental supply voltage will contain components at $2s\omega_{se}$ and $2(1 - s)\omega_{se}$.

An example of this was simulated by Trzynadlowski *et al.* [42], as shown in Figure 9.15, for a six-pole induction motor with a damaged rotor fed at $f_{se} = 60$ Hz with a very large slip at this operating condition of 15.8 per cent, therefore the side bands on the current PSD spectrum due to damage are positioned above and below the 60 Hz fundamental, at 41 Hz and 79 Hz, whereas on the power PSD spectrum the damage side band is positioned at 19 Hz.

9.5.5 Shaft voltage or current

Many electrical power utilities have attempted to monitor the voltages induced along the shafts of electrical machines in the hope that they may be a useful indicator of machine core or winding degradation and because they can give rise to large shaft currents, which are damaging to bearings. Figure 9.16 shows how a voltage can be induced between contacts sliding on a rotating machine shaft whenever fluxes in the machine are distorted, either from the normal radial and circumferential pattern, or from the normal axial pattern. These sliding contacts may be the result of rubs on defective bearings or seals or could be brushes placed to detect flux distortion. The brushes would normally be placed at either end of the machine to embrace the complete shaft flux circuit. If a fault, such as a rotor winding shorted turn, produces a rotating distortion of the field in the radial and circumferential plane then an AC or pulsating shaft voltage results. If a fault produces a distortion of the field in the axial direction then this gives rise in effect to a homopolar flux that produces a DC shaft voltage. In steam turbine-driven machines, shaft voltages can also be produced by electrostatic action, where the impingement of water droplets on turbine blades charges the shaft. Verma and Girgis [44] have given a full report on the mechanisms for

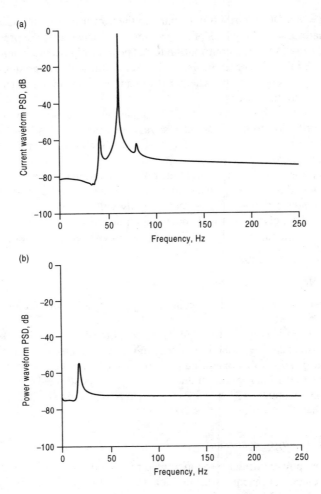

Figure 9.15   Spectra from a simulation of an induction motor with broken rotor bars. The upper curve shows the current spectrum with the typical fault side bands. The lower graph shows the power spectrum with the fault side band shifted down to DC. (a) Current spectrum with typical fault sidebands; (b) power spectrum with fault sideband shifted down to DC. [Taken from Trzynadlowski et al. [42]. © IEEE (1999)]

the production of shaft voltages and currents and the faults they may indicate. Methods of monitoring shaft voltages usually include making AC and DC measurements of the voltage and sometimes analysing the harmonic content of the waveform. Verma proposed a comprehensive shaft voltage monitor in [44] and Nippes a more up-to-date version in [45].

Our experience, however, is that shaft voltage has not proved to be a useful parameter for continuous monitoring. The voltage is difficult to measure continuously,

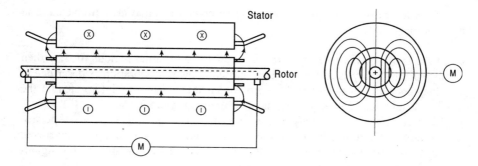

Figure 9.16   The production of shaft voltages due to asymmetries in the magnetic field of the machine

because of the unreliability of shaft brushes, particularly when they are carrying only a small measurement current. In addition, it has been shown by the authors' colleagues that any damage to the core and winding would need to be substantial before a significant variation in shaft voltage occurred. This should not detract, however, from the normal regular maintenance procedure of measuring shaft voltages at bearing pedestals, from time-to-time, in order to check the pedestal insulation and to confirm that there is no tendency for large shaft currents to flow.

9.5.6   Mechanical and electrical interaction

There is considerable commonality between the approaches described in this section and those of Chapter 8, concerned respectively with the electrical and mechanical responses to faults in the electrical machine. This commonality arises from the airgap magnetic field and it will be important for future condition-monitoring strategies on electrical machines to have a clear understanding of the link between the electrical and mechanical root causes of faults and their effect on the airgap field.

Mechanical engineers will tend to look for faults via the vibration spectrum and electrical engineers will look to the current, flux or power spectra. The reality is that all these spectra are closely coupled as set out in Chapter 8 and it may be better to look for some faults via the vibration signal and for others via the flux, voltage, current, power or vibration signals. In Chapter 8, Table 8.6 we gave a summary of the lateral vibration angular frequencies caused by various faults. In this chapter, Tables 9.1 and 9.2, we give a similar summary respectively of current and flux angular frequencies caused by various faults.

9.6   Effects of variable speed operation

Variable speed converters are being applied in increasing numbers to electrical machines. While this book has been prepared on the basis that it does not deal with the specific problems of condition-monitoring electrical machines driven at variable speed, it will be useful here to identify the key problems that may arise from condition monitoring such drives.

- If the speed of the drive remains constant for substantial periods of time then the spectral analysis of flux, voltage, current, power or vibration as described in Chapters 8 and 9 can still be done provided that the results are interpreted for the speed and base frequency when the measurements were made.
- However, if the speed is varying then non-stationary techniques, such as spectrograms over short time intervals, wavelets or Wigner–Ville techniques, dependent on the rate of change of speed, will need to be used with the monitoring signals because the spectra will not be stationary.
- If speed is varying under control loop action then the frequency content of the monitoring signals will be affected by the bandwidth of the controller, as described by Bellini *et al.* [46]. In this case it is possible for the drive controller to suppress fault harmonic signals in the flux, voltage, current, power or vibration. However, Bellini has shown that it is still possible to extract condition-monitoring information from signals derived inside the controller; in Bellini's example, this was done from the direct axis current, i_d.
- When supplied from a variable speed drive all terminal quantities of the electrical machine, axial flux, current and power, will be polluted by harmonics generated by the drive and filtering this pollution will be essential to obtaining a good signal-to-noise ratio with condition-monitoring signals from such a machine.

Chen *et al.* [47] has demonstrated a condition-monitoring system for variable speed induction motors using the power line as his communication channel. This is a brave attempt to achieve universal monitoring and wisely the author has avoided monitoring the electrical signals but concentrated upon monitoring the winding temperature.

Table 9.1 *Axial flux angular frequency components related to specific induction machine assymmetries equation (9.22). [Taken from Vas [48]]*

	Space harmonic of the stator winding MMF			
	$k = 1$	$k = 3$	$k = 5$	$k = 7$
	Angular frequency components, rad/s			
Stator asymmetry	$s\omega_{se}$ $(2 - s)\omega_{se}$	$(3s - 2)\omega_{se}$ $(4 - 3s)\omega_{se}$	$(5s - 4)\omega_{se}$ $(6 - 5s)\omega_{se}$	$(7s - 6)\omega_{se}$ $(8 - 7s)\omega_{se}$
Rotor asymmetry	$s\omega_{se}$ $(2s - 1)\omega_{se}$	$(3 - 2s)\omega_{se}$ $(4s - 3)\omega_{se}$	$(5 - 4s)\omega_{se}$ $(6s - 5)\omega_{se}$	$(7 - 6s)\omega_{se}$ $(8s - 7)\omega_{se}$

Table 9.2 *Electrical angular frequency components related to specific electrical machine faults*

	Angular frequency components, rad/s		
	Current	Flux	Power
Mechanical faults	Oil whirl and whip in sleeve bearings	$(0.43 \text{ to } 0.48)\omega_{se}/p$		
	Unbalanced mass on rotor of a synchronous machine		ω_{se}/p	
	Dynamic eccentricity in a synchronous machine	$2\omega_{se}P$	$2\omega_{se}/p$	
	Dynamic displacement of shaft in bearing housing of a synchronous machine		$\omega_{se}/p, 2\omega_{se}/p \cdots$	
	Static misalignment of rotor shaft in a synchronous machine	$\omega_{se}/p, 2\omega_{se}/p,$ $3\omega_{se}/p \cdots$	$\omega_{se}/p, 2\omega_{se}/p,$ $3\omega_{se}/p \cdots$	
	Static and dynamic eccentricity in induction machine	$\omega_{se}\left[(nN_r \pm k_e)\dfrac{(1-s)}{p} \pm k\right]$	$\omega_{se}\left[k_e\dfrac{(1-s)}{p} \pm k\right]$	
Electrical faults	Broken rotor bar in induction machine	$(1 \pm 2ns)\omega_{se}/p$	$s\omega_{se}\cdots$ $(2s-1)\omega_{se}\cdots$	$2ns\omega_{se}/p$ $2n(1-s)\omega_{se}/p$
	Stator winding faults in a synchronous machine	$2\omega_{se}, 4\omega_{se}\cdots$	ω_{se}	$\omega_{se}, 3\omega_{se}$
	Stator winding faults in an induction machine		$s\omega_{se}\cdots$ $(2-s)\omega_{se}\cdots$	

Definitions: ω_{sm} = mechanical vibration frequency on the stator side; ω_{se} = electrical supply frequency on the stator side; ω_{se} = electrical supply frequency, Hz; N_r = integer number of rotor slots; N = rotational speed in rev/min; $f\omega_{rm} = \frac{2\pi N}{60}$ mechanical rotational frequency, Hz; $\omega_{rm} = \omega_{se}/p$ for a synchronous machine; $\omega_{rm} = (1-s)\omega_{se}/p$ for an asynchronous machine rotor speed slip, $0-1$; p = pole pairs; n = an integer; k_e = eccentricity order, zero for static eccentricity, low integer value 1, 2, 3 ... for dynamic eccentricity; k = space harmonic of the stator winding MMF, 1, 3, 5, 7 ...

9.7   Conclusion

This chapter has shown that electrical techniques are powerful tools for the condition monitoring of electrical machines, particularly axial leakage flux, current and power, offering the potential to provide a general condition-monitoring signal for the machine.

The availability of high-quality, digitally sampled mechanical vibration and electrical terminal data from electrical machines opens the possibility for more comprehensive monitoring of the machine and prime mover or driven machine combinations. However, these signals generally require broad bandwidth (>50 kHz) and a high data rate for adequate analysis. Therefore the principal difficulty of applying these techniques is the complexity of the necessary spectral analysis and interpretation of their content.

This situation is made more difficult if variable speed drives are involved because time domain signals may no longer be stationary and will also be polluted by harmonics from the power electronic drive.

Comprehensive monitoring of an electrical machine can be achieved by measuring shaft flux, current, power and electrical discharge activity. These are broad bandwidth (generally >50 kHz) signals requiring complex analysis. Shaft flux, current and power signals are capable of detecting faults in both the electrical and mechanical parts of a drive train. Shaft voltage or current is an ineffective condition-monitoring technique for electrical machines. Shaft flux monitoring is non-invasive and uses a single sensor but it is complex to analyse and untested in the field. Current monitoring is also non-invasive, but uses existing sensors and has established itself as motor current spectral analysis, a reliable and widely accepted technique for machine monitoring. Power monitoring is also non-invasive, uses existing sensors but requires less bandwidth (<10 kHz) and less complex spectral interpretation to detect faults but is not yet widely accepted, so it deserves investigation for future development.

9.8   References

1. Rai R.B. *Airgap Eccentricity in Induction Motors*. ERA Report, 1974: 1174–88.
2. Dorrell D.G., Thomson W.T. and Roach S. Analysis of airgap flux, current, and vibration signals as a function of the combination of static and dynamic airgap eccentricity in 3-phase induction motors. *IEEE Transactions on Industry Applications* 1997; **IA-33**: 24–34.
3. Hargis C. Steady-state analysis of 3-phase cage motors with rotor-bar and end-ring faults. *IEE Proceedings, Part B, Electric Power Applications* 1983; **130**: 225.
4. Nandi S., Bharadwaj R.M., Toliyat H.A. and Parlos A.G. Performance analysis of a three-phase induction motor under mixed eccentricity condition. *IEEE Transactions on Energy Conversion* 2002; **EC-17**: 392–9.

5. Penman J., Sedding H.G., Lloyd B.A. and Fin W.T. Detection and location of interturn short circuits in the stator windings of operating motors. *IEEE Transactions on Energy Conversion* 1994; **EC-9**: 652–8.

6. Penman J. and Stavrou A. Broken rotor bars: their effect on the transient performance of induction machines. *IEE Proceedings, Part B, Electric Power Applications* 1996; **143**: 449–57.

7. Thomson W.T. and Fenger M. Current signature analysis to detect induction motor faults. *IEEE Industry Applications Magazine* 2001; **7**: 26–34.

8. Thomson W.T. and Barbour A. On-line current monitoring and application of a finite element method to predict the level of static airgap eccentricity in three-phase induction motors. *IEEE Transactions on Energy Conversion* 1998; **EC-13**: 347–57.

9. Thomson W.T., Rankin D. and Dorrell D.G. On-line current monitoring to diagnose airgap eccentricity in large three-phase induction motors-industrial case histories verify the predictions. *IEEE Transactions on Energy Conversion* 1999; **EC-14**: 1372–8.

10. Michiguchi Y., Tonisaka S., Izumi S., Watanabe T. and Miyashita I. Development of a collector ring monitor for sparking detection on generators. *IEEE Transactions of Power Apparatus and Systems* 1983; **PAS-102**: 928–33.

11. Warrington A.W.V.C. *Protective Relays, their Theory and Practice, Vol. 1.* London: Chapman and Hall; 1982.

12. Khudabashev, K.A. Effect of turn short-circuits in a turbo generator rotor on its state of vibration. *Elekt Stantsii USSR* 1961; **7**: 40–5.

13. Albright D.R. Inter-turn short circuit detector for turbine generator rotor windings. *IEEE Transactions Power Apparatus and Systems* 1971; **PAS-50**: 478–83.

14. Kryukhin S.S. A new principle for synchronous machine protection from rotor winding inter-turn and double earth faults. *Elect. Technol. USSR* 1972; **2**: 47–59.

15. Pöyhönen S., Negrea M., Jover P., Arkkio A. and Hyötyniemi H. Numerical magnetic field analysis and signal processing for fault diagnostics of electrical machines. *COMPEL: The International Journal for Computation and Mathematics in Electrical and Electronic Engineering* 2003; **22**: 969–81.

16. Ramirez-Nino J. and Pascacio A. (2001). Detecting interturn short circuits in rotor windings. *IEEE Transactions on Computer Applications in Power* 2001; **CAP-14**: 39–42.

17. Streifel R.J., Marks II R.J., El-Sharkawi M.A. and Kerszenbaum I. Detection of shorted-turns in the field winding of turbine-generator rotors using novelty detectors – development and field test. *IEEE Transactions on Energy Conversion* 1996; **EC-11**: 312–17.

18. Tavner P.J., Gaydon B.G. and Ward D.M. Monitoring generators and large motors. *IEE Proceedings, Part B, Electric Power Applications* 1986; **133**: 169–80.

19. Kamerbeek E.M.H. Torque measurements on induction motors using Hall generators or measuring windings. *Philips Technical Review* 1974; **34**: 152–62.

20. Seinsch H.O. Detection and diagnosis of abnormal operating conditions and/or faults in rotating electrical machines. *Schorch Berichte* 1986.

21. Jufer M. and Abdulaziz M. Influence d'une rupture de barre ou d'un anneau sur les characteristiques externes d'un moteur asynchrone a cage. Bulletin SEV/VSE, Switzerland, 1978; **69**.

22. Menacer A., Nat-Said M.S., Benakcha A.H. and Drid S. Stator current analysis of incipient fault into induction machine rotor bars. *Journal of Electrical Engineering* 2004; **55**: 122–30.

23. Li W.D. and Mechefske C.K. Detection of induction motor faults: A comparison of stator current, vibration and acoustic methods. *Journal of Vibration and Control* 2006; **12**: 165–88.

24. Rickson C.D. Protecting motors from overload due to asymmetrical fault conditions. *Electrical Review* 1983: 778–80.

25. Trutt F.C., Sottile J. and Kohler J.L. Online condition monitoring of induction motors. *IEEE Transactions on Industry Applications* 2002; **IA-38**: 1627–32.

26. Kohler J.L., Sottile J. and Trutt F.C. Condition monitoring of stator windings in induction motors. I. Experimental investigation of the effective negative-sequence impedance detector. *IEEE Transactions on Industry Applications* 2002; **IA-38**: 1447–53.

27. Sottile J., Trutt F.C. and Kohler J.L. Condition monitoring of stator windings in induction motors. II. Experimental investigation of voltage mismatch detectors. *IEEE Transactions on Industry Applications* 2002; **IA-38**: 1454–59.

28. Sottile J., Trutt F.C. and Leedy A.W. Condition monitoring of brushless three-phase synchronous generators with stator winding or rotor circuit deterioration. *IEEE Transactions on Industry Applications* 2006; **IA-42**: 1209–15.

29. Tavner P.J. Permeabilitatsschwankungen auf Grund von Walzeffekten in Kernplatten ohne Valenzrichtung. *Elektrotechnik und Maschinenbau* 1980; **97**: 383–6.

30. Jordan H. and Tagen F. Wellenflusse infolge con schwankungen des luftapalteitwertes. *Elektrotechnik und Zeitschrift* 1964; **85**: 865–7.

31. Jordan H., Kovacs K. and Roder P. Messungen des schlupfes von asynchronmaschinen mit einer spule. *Elektrotechnik und Zeitschrift* 1965; **86**: 294–6.

32. Erlicki M.S., Porat Y. and Alexandrovitz A. (1971). Leakage field changes of an induction motor as indication of non-symmetric supply. *IEEE Transactions on General Applications* 1971; **GA-7**: 713–7.

33. Penman J., Hadwick G. and Stronach A.F. Protection strategy against faults in electrical machines. Presented at the International Conference on Developments in Power System Protection, London, 1980.

34. Yacamini R., Smith K.S. and Ran L. Monitoring torsional vibrations of electromechanical systems using stator currents. *Journal of Vibration and Acoustics* 1998; **120**: 72–9.

35. Ran L., Yacamini R. and Smith K.S. Torsional vibrations in electrical induction motor drives during start up. *Journal of Vibration and Acoustics* 1996; **118**: 242–51.

36. Joksimovic G.M. and Penman J. The detection of inter-turn short circuits in the stator windings of operating motors. *IEEE Transactions on Industrial Electronics* 2000; **IA-47**: 1078–84.

37. Stavrou A., Sedding H.G. and Penman H. Current monitoring for detecting inter-turn short circuits in induction motors. *IEEE Transactions on Energy Conversion* 2001; **EC-16**: 32–7.

38. Bellini A., Filippetti F. et al. Quantitative evaluation of induction motor broken bars by means of electrical signature analysis. *IEEE Transactions on Industry Applications* 2001; **IA-37**: 1248–55.

39. Bellini A., Filippetti F., Franceschini G., Tassoni C. and Kliman G.B. On-field experience with online diagnosis of large induction motors cage failures using MCSA. *IEEE Transactions on Industry Applications* 2001; **IA-38**: 1045–53.

40. Henao H., Demian C. and Capolino G.A. A frequency-domain detection of stator winding faults in induction machines using an external flux sensor. *IEEE Transactions on Industry Applications* 2003; **IA-39**: 1272–9.

41. Henao H., Razik H. and Capolino G.A. Analytical approach of the stator current frequency harmonics computation for detection of induction machine rotor faults. *IEEE Transactions on Industry Applications* 2005; **IA-41**: 801–7.

42. Trzynadlowski A.M., Ghassemzadeh M. and Legowski S.F. Diagnostics of mechanical abnormalities in induction motors using instantaneous electric power. *IEEE Transactions on Energy Conversion* 1999; **EC-14**: 1417–23.

43. Hunter D.J., Tavner P.J., Ward D.M. and Benaragma D.. Measurements of the harmonic components of the instantaneous electrical power delivered at the terminals of a 500 MW turbogenerator. Presented at the 3rd International Conference on Sources and Effects of Power System Disturbances, London, 1982.

44. Verma S.P. and Girgis R.S. *Shaft Potentials and Currents in Large Turbogenerators*. Report for the Canadian Electrical Association, 1981.

45. Nippes P.I. Early warning of developing problems in rotating machinery as provided by monitoring shaft voltages and grounding currents. *IEEE Transactions on Energy Conversion* 2004; **EC-19**: 340–45.

46. Bellini A., Filippetti F., Franceschini G. and Tassoni C. Closed-loop control impact on the diagnosis of induction motors faults. *IEEE Transactions on Industry Applications* 2000; **IA-36**: 1318–29.

47. Chen S., Zhong E. and Lipo T.I. A new approach to motor condition monitoring in induction motor drives. *IEEE Transactions on Industry Applications* 1994; **IA-30**: 905–11.

48. Vas P. *Parameter Estimation Condition Monitoring and Diagnosis of Electrical Machines*. Oxford, UK: Clarendon Press; 1996.

Chapter 10
Electrical techniques: discharge monitoring

10.1 Introduction

Chapter 9 concentrated on the perturbations to current, flux and power at the terminals due to faults in the machine. However, there is another effect at work in high-voltage electrical machines and that is perturbations to the voltage and current wave fed to the machine as a result of electrical disturbances or discharges in the insulation system.

This chapter deals with a more specialised terminal analysis than considered in Chapter 9, which has the potential to detect those discharge activities present in the machine winding insulation, particularly in high voltage machines. One of the reasons this technique has received so much attention is because the insulation system lies at the heart of every electrical machine, as described in Chapter 2, and its deterioration can be relatively slow, as described in Chapter 3. Therefore it should be a good target for condition monitoring.

10.2 Background to discharge detection

Discharge behaviour is complex and can be categorised in ascending order of energy and damage as

- corona discharge,
- partial discharge,
- spark discharge,
- arc discharge.

A well-made insulation system will exhibit low-level corona discharge on the surface of the insulation at AC voltages above 4 kV rms to ground. If there are voids inside the body of the insulation system those voids will also exhibit partial discharges at the points in the voltage cycle when the local electric field strength exceeds the Paschen curve level for the gas in the void at that temperature. Neither that surface or body activity is necessarily damaging unless the activity is sufficiently powerful to degrade the insulation system, as described in Chapter 7 Section 7.2. This activity can progressively worsen depending on the quality of the insulation, the local strength of the field and the mechanical and electrical conditions to which the material is subjected. There are certain parts of high-voltage winding insulation systems that are particularly vulnerable to discharge activity, namely:

- the stator slot wall where partial discharge activity can erode and damage the main wall insulation;

- the slot emergence where coils emerge from the earth protection of the slot and the insulation system is exposed to surface discharge activity;
- the end-winding surfaces, which can be subjected to damaging discharge activity, particularly at the phase separation regions, when the surface is wet or dirty or both.

A study of the failure mechanisms in Chapter 2 shows that electrical discharge activity is an early indicator of many electrical faults in machine stators and the activity is also related to the remaining life of the insulation system. The accurate detection of discharge activity could therefore give valuable early warning of failure and could provide information about the remaining life of the insulation.

Electrical discharges are transitory, low-energy disturbances that radiate electromagnetic, acoustic and thermal energy from the discharge site. That conducted energy causes perturbations to the waveforms of the voltage and current both within the machine and at the machine terminals. The earliest applications of partial discharge detection were to isolated insulation components, such as transformer and high voltage machine bushings or most successfully to the stop joints in oil-cooled EHV cables [1], where the insulation under inspection is close to the coupling circuit and is energised solely at the phase voltage of the cable. After successful application on cable stop joints the method was then applied to the windings of high-voltage turbine and hydro generators. However, an electrical machine winding represents a much more complex insulation system than a stop joint, see Figure 10.1.

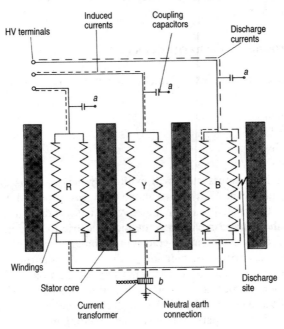

Figure 10.1   The complex structure of a typical turbine generator stator winding and its arrangement [Taken from Tavner and Jackson [2]]

For example

- The insulation is distributed throughout the length of the winding, which may represent many hundreds of meters in length.
- It is energised with a distributed potential, applying full line voltage between the winding ends but zero voltage at the neutral point.
- The winding is complex from a propagation point of view:
  - Containing propagation path bifurcations
    - at coil connections,
    - where sections of the winding are connected between phases,
    - where individual phases consist of parallel paths.
  - It also contains many surge impedance transitions throughout its length (see Figure 10.2)
    - as it enters and emerges from the slot,
    - as the coupling varies between winding components, in slot and end-winding portions,
    - as it connects to the main terminals of the machine.

The following description traces the development of electrical machine winding partial discharge detection up to the present day and demonstrates the value of discharge detection techniques.

10.3 Early discharge detection methods

10.3.1 RF coupling method

The earliest work, summarised by Harrold *et al.* [3], describes a technique developed by Westinghouse in the US to detect on-line the presence of subconductor arcing in the stator windings of large steam turbine generators, by measuring perturbations in the winding current. Arcing activity produces very wide-band electromagnetic energy, some of which propagates into the neutral connection of the star-connected winding. Emery used a ferrite-cored RF current transformer (RFCT) wrapped around the neutral cable to couple to this activity, which he detected using a quasi-peak, radio interference field intensity (RIFI) meter, as shown in Figure 10.3. The neutral cable was chosen as a good measurement location because it has a low potential with respect to ground and because arcing at any location in the generator causes RF current to flow into the neutral lead.

The RFCT has a frequency response from 30 Hz to 30 MHz and the RIFI meter has a narrow bandwidth of $\sim$10 kHz centred at about 1 MHz. The centre frequency is nominally tuned to match resonances in the winding that the arcing activity excites; the RIFI meter effectively measures the average peak energy received by the instrument. The RFCT and RIFI meters are proprietary items and a simple monitoring system can be assembled using these components. Westinghouse have also developed a specialised RF monitor based on this technique [4]. The monitor interfaces with a remote panel located in the machine control room and provides a permanent record of arcing activity, with alarm indications to the operator when a severe increase occurs.

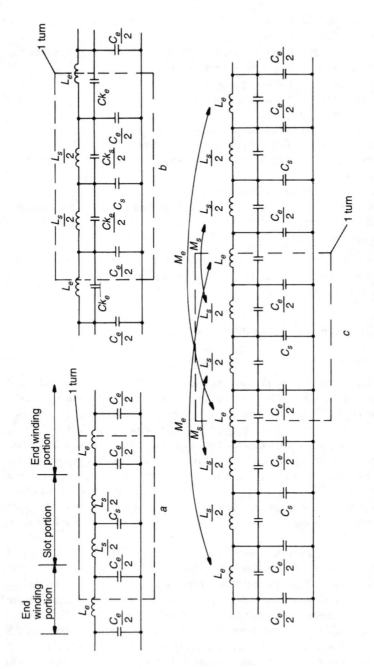

Figure 10.2 The electrical structure of one turn of a typical stator winding and the complex coupling arrangements. (a) Simplest model, one turn represented by a Tee equivalent circuit. (b) Increased complexity, one turn represented by a series of Tee equivalent circuits. (c) Further complexity, one turn represented by a series of Tee equivalent circuits, with the core and end winding couplings approximated. [Taken from Tavner et al. [2]]

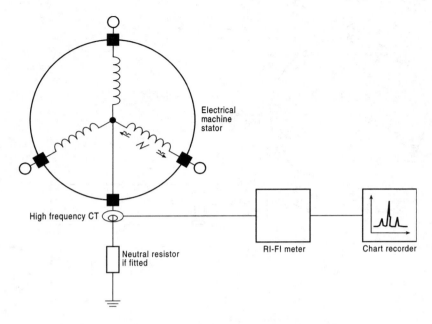

Figure 10.3   Detection of discharge activity in a generator [Taken from Emery et al.
[4]. © IEEE (1981)]

Harrold *et al*. had shown in [3], with further details by Emery *et al*. [4] and Harrold and Emery [5], positive proof of the detection, not only of subconductor arcing, but also of sparking in other parts of steam-turbine-driven generators. However, the change in signal level when arcing occurs, from the RIFI meter tuned to 1 MHz, is not dramatic. An increase of less than 50 per cent of the unfaulted indication is typical and this makes the setting of alarm levels for such a monitor difficult.

Timperley [6] has used this technique and applied it not only to steam turbine generators but also to hydroelectric machines in the American Electric Power service. He appears to have used wider bandwidth quasi-peak RIFI instruments connected to the neutral RFCT because he analyses the received signal in both the frequency and time domains. Using the technique he has shown some evidence of the detection of slot discharge activity on hydroelectric machines and other forms of unexpected corona activity.

10.3.2   Earth loop transient method

Wilson *et al*. [1], in the UK, devised a similar technique to that of Emery as a cheap method of detecting discharge activity on-line in a wide range of high-voltage plant. Initially it was applied to a relatively small, identifiable section of insulation, such as a stop joint in an oil-insulated EHV cable, and was known as an earth loop transient monitor. It has since been applied to the insulation of generator and motor stator windings and aims to look for partial discharge activity in the bulk of the winding.

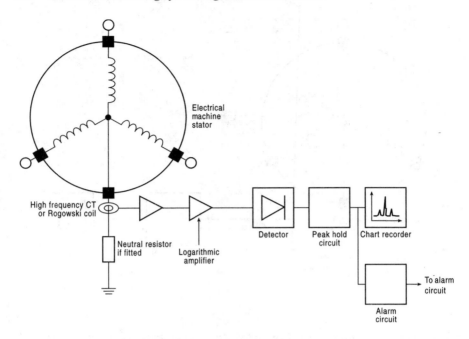

Figure 10.4   Continuous monitor of discharge activity using an earth loop transient monitor [Taken from Wilson et al. [1]]

It uses a Rogowski coil, wrapped around the neutral cable of the machine winding, and the detector is a narrow-band instrument that measures the average peak energy received by the instrument as shown in Figure 10.4. The Rogowski coil is an air-cored solenoidal search coil that is closed on itself round a current-carrying conductor. The manufacture of these coils has been patented in the UK by Ward and Exon [7] and they are called Rogowski coils to distinguish them from their ancestor, the Chattock potentiometer, which is not a wraparound coil as needed for current measurements. The frequency response of the Rogowski coil is relatively wide but the detector has a narrow bandwidth of ∼15 kHz centred at a value determined by the application and the background radio noise, but usually for a generator or motor winding this would be 1 MHz.

The monitor is calibrated in picocoulombs and is provided with alarm circuits, so that when the discharge level exceeds a warning threshold an alarm signal can be transmitted to the plant control room. Wilson has explained that when applying this technique to a distributed insulation system, such as a machine winding, care must be taken in the calibration, because energy may be propagated to the instrument from a number of different discharge sites in the insulation simultaneously. Geary *et al.* [8] provide a theoretical model for the manner in which energy, in the frequency band detected by the instrument, is propagated from the discharge site in to the winding neutral and he has shown how this propagation depends critically on the configuration of the winding and the size of the stator core, as described in Section 10.2.

10.3.3 Capacitive coupling method

An alternative technique, where perturbations in voltage waveforms are detected at the machine terminals, has been described by Kurtz and Stone [9] and applied primarily to hydroelectric alternators in Canada aimed at detecting the slot discharge activity with which these high-voltage air-cooled windings are afflicted. Connection to the winding is made through coupling capacitors connected to the line terminals of the machine as shown in Figure 10.5. Discharge pulses are coupled through these capacitors to a specialised pulse height analyser, which characterises, in the electronics in the analyser is 80 MHz and this is considered sufficient to capture partial discharge pulses with rise times of the order of 1–10 ns. In the early days of this technique the coupling capacitors had to be connected to the machine during an outage but in a later publication Kurtz *et al.* [10] describe how capacitative couplers can be permanently built into the phase rings of the machine so that the measurements can be made without service interruptions. In addition these permanent couplers are also intended to ensure that discharge activity from the electrical supply system, to which the machine is connected, is rejected. However, the pulse height analysis of discharges by this method is still carried out at intervals during the life of the machine rather than continuously on-line. Kurtz has not shown how the electromagnetic energy from a discharge site is propagated through the winding to the coupling capacitors but has

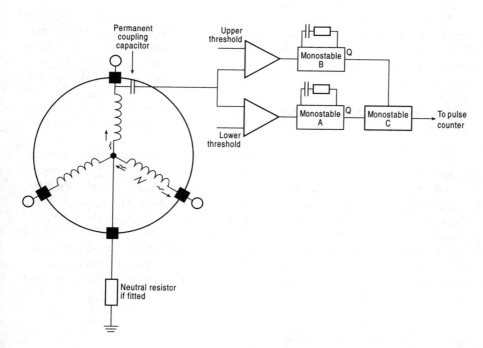

Figure 10.5   Detection of discharge activity using a coupling capacitor [Taken from Kurtz et al. [10]]

shown empirically that the method is capable of detecting slot discharges and the steady deterioration of winding insulation with time.

10.3.4   Wideband RF method

All the techniques described earlier operate at relatively low frequencies (1–80 MHz) and detect the electromagnetic energy propagated along the winding to the neutral or line end connections [2].

In any healthy machine there will be a background of corona and partial discharge activity that will vary from machine to machine and also vary with time. Malik *et al.* [11] showed that damaging discharge pulses, such as serious partial discharges, sparking or arcing, have faster rise-times than the background corona and partial discharge activity and therefore produce a much higher frequency of electromagnetic energy (350 MHz). Malik *et al.* showed that if this energy is detected, at as high a frequency as possible, the ratio of damaging discharge signals to background activity is increased. Frequencies of electromagnetic energy >4 MHz propagate from the discharge site by radiation from the winding, not by propagation along the winding as in the case with the lower-frequency techniques. This radiation can be detected by an RF aerial located either inside the enclosure of the machine or outside, close to an aperture in it. This technique is very similar to that proposed by Michiguchi *et al.* [12] for detecting faults in brushgear.

Malik *et al.* [11] described a monitor for detecting damaging discharge activity in a turbine generator stator winding based on this idea, as is shown in Figure 10.6. It receives the energy from the aerial and amplifies it before detection. The monitor contains a band-pass filter that is tuned above the cut-off frequency of background activity (~350 MHz), avoiding any interference from nearby radio or radar stations, to the part of the spectrum that is of interest. The output of the monitor is a chart record that shows the instants in time at which the energy due to damaging discharge activity exceeded a threshold value. This threshold can be set according to the level of background activity in the machine.

The monitor has been used successfully on large operational steam turbine-driven generators, where the aerial is relatively easy to fit outside the casing by mounting it close to the neutral point connection. The instrument has positively identified proven subconductor arcing and has been shown to detect other forms of damaging discharge activity. Its advantage over the other techniques is that by detecting at higher frequencies the signal-to-noise ratio of damaging activity to background is much larger and this makes it much easier to determine alarm settings for the instrument, as is shown in Figure 10.7. On the other hand the received signal cannot be related directly to discharge magnitudes in picocoulombs and all timing information about the discharge is lost.

10.3.5   Insulation remanent life

Besides identifying specific faults, partial discharge detection and measurement have also held out the possibility of determining the remanent life of insulation systems

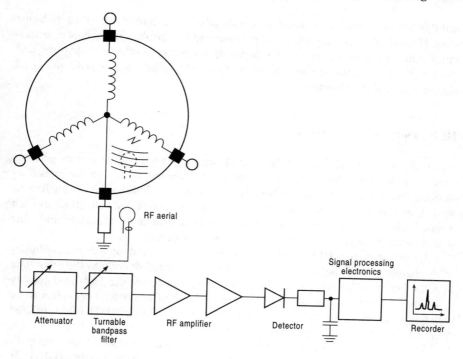

Figure 10.6   *Detection of damaging discharge activity using radiofrequency energy [Taken from Malik et al. [11]]*

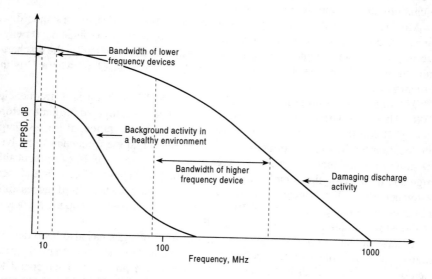

Figure 10.7   *Radiofrequency energy from background and damaging discharge activity*

and this has occupied a considerable amount of the literature as described by Stone *et al.* [13,14]. This process requires measurements of insulation resistance, polarisation index, capacitance, dissipation factor tip-up, partial discharge magnitude, and discharge inception voltage as well as on-line partial discharge activity and is an elusive and difficult objective.

10.4 Detection problems

Lower–frequency work (1–80 MHz), involving propagation along a homogeneous conductor, suggests that a discharge site can be located by timing the arrival of discharge pulses received at a number of different sensors. This has now been demonstrated, as shown in the next section on modern techniques, but reflections at each discontinuity in the winding or variation in its insulation and connection make the identification of pulses at the terminals extremely difficult.

The low-frequency RFCT devices produced an output calibrated in picocoulombs of discharge activity, because the response can be related directly to the amplitude of a discharge calibration pulse. This allows the user to see the measured activity in discharge terms so that he can decide what level of activity he considers to be damaging. Although the Rogowski coil is easier to fit than the RFCT it has suffered from its very low sensitivity and is now rarely used.

Various other partial discharge detectors have been investigated, including the portable Tennessee Valley Authority (TVA) probe and Discharge Locator (DL) [15], and the Stator Slot Coupler (SSC) [16], which can be mounted above the stator conductor in the slot beneath the wedge (see Figure 2.4). These confer some advantages in enabling the operator to locate discharge activity.

More recently, partial discharge measurement has been made using a standardised capacitive coupler, based on the Canadian experience, which is a robust power engineering component of capacitance C that can easily be calibrated in Q_m, the discharge value in millivolts across the coupler, where Q_m in millivolts Q/C, where Q is the discharge value.

The higher signal-to-noise potential of wideband RF techniques at frequencies (100 MHz–1 GHz) described in Section 10.3.4 has been investigated further by Stone *et al.* [17] and from this work there would appear to be the potential for discharge site location by the use of a directional aerial. However, at the frequencies involved the dimensions of such an aerial, diameter from 0.6 to 6 m, would be impracticably large and in any case the complex machine structure will cause reflections and local resonances that will disrupt location. However, this method has been used successfully in transformers using three or more RF sensors located in positions on the surface of the transformer tank to triangulate the location of defects.

Ultrasonic detection of the noise emitted from discharges has also been demonstrated on transformers to be much more effective at fault location than electrical methods in rotating electrical machines. So despite the potential of wideband RF techniques their lack of location ability means that the technique has now largely been superseded.

It must be made clear at this stage that there is less value in being able to detect discharge activity if, when the machine is taken out of service, that activity is impossible to locate. It would be preferable to allow damaging discharge activity to continue until it had reached such a pitch that the damage was observable. This highlights the problem of what constitutes a significant level discharge activity.

In all partial discharge detection systems it is necessary to shield the desired partial discharge activity signal from external noise, either due to partial discharge activity in the connections and switchgear of the electrical machine or due to sparking in brushgear or harmonic activity due to nearby power electronics. In some cases, the electrical machine is connected to the system via a lengthy, low-discharge cross-linked polyethylene cable, that provides the necessary filter, but it is not always so. These noise and calibration issues are dealt with by Geary *et al.* [8], Wood *et al.* [18], Campbell *et al.* [19], Stone and Sedding [20] and Kemp *et al.* [21].

Another fundamental problem for all partial discharge detection systems is that the discharge activity of identical windings in different machines exhibit large variations in background activity, due to variations in ambient conditions, small changes in the insulation homogeneity and noise conditions. Therefore one cannot say with any certainty what the background discharge activity for a winding should be, and this activity will vary naturally with time regardless of whether any damaging activity is taking place. An example of partial discharge detection results taken from a database from a number of 13.8 kV turbine generators in the US, measured using capacitive couplers [22], are shown in Table 10.1, where the variations with voltage and between machines can be seen.

Table 10.1 q_m *in millivolts from a database of results on a set of 13.8 kV turbine generators, measured using 80 pf capacitive couplers. [Taken from Stone et al. [13]]*

Rated V	2–4 kV	6–8 kV	10–12 kV	13–15 kV	16–18 kV	>19 kV
25%	2	6	27	9	145	120
50%	15	29	63	79	269	208
75%	57	68	124	180	498	411
90%	120	247	236	362	1 024	912
Average	89	88	121	168	457	401
Maximum	2 461	1 900	3 410	3 396	3 548	3 552

10.5 Modern discharge detection methods

Modern on-line discharge detection methods for rotating electrical machines have developed from the early work described in Section 10.3, improved to resolve some of the problems described in Section 10.4 and can now be divided into two techniques.

- The hydro generator partial discharge analyser (PDA), based on Section 10.3.3 (p. 235) (see Figure 10.8 and Lyles and Goodeve [23]).

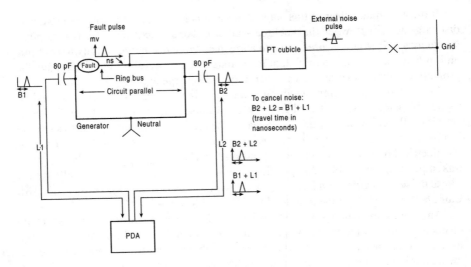

Figure 10.8   Diagram of the coupler connections and electronics for the partial discharge analyser (PDA) [Taken from Stone et al. [22]. © IEEE (2004)]

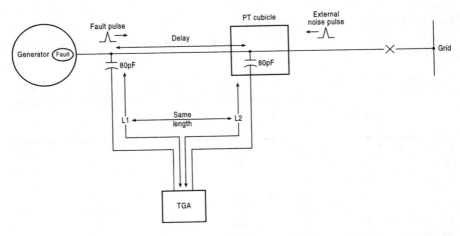

Figure 10.9   Diagram of the connections and electronics for the turbine generator analyser (TGA) [Taken from Stone et al. [22]. © IEEE (2004)]

- The motor or turbine generator analayser (TGA), based on Sections 10.3.1 and 10.3.2 (p. 235), which adopts a variant of the PDA to improve the signal-to-noise ratio of partial discharge detection on windings of the types found in motors and turbine generators (see Figure 10.9). Experience with this technique is described in Stone *et al.* [24].

An IEEE standard has also now been devised [25] to standardise the measurement of partial discharge on rotating machines.

Work is continuing on partial discharge detection methods, including work on lower-voltage, 4 kV motors by Tetrault *et al.* [26], work on large machines Kheirmand *et al.* [27] and an example of an insulation monitoring system that detects insulation leakage currents by measuring machine terminal voltages on an induction motor by Lee *et al.* [28].

10.6   Conclusion

Discharge measurement has shown itself to be the most problematic electrical method of condition monitoring. It requires special sensors, wide bandwidth (> 100 kHz) and very complex analysis for fault detection and can only be recommended where a specific high-voltage failure mode is being searched for in a known location on a large machine. It addresses one of the most vital parts of the electrical machine, it can detect global effects, including possibly remanent life of the machine insulation, and it does give a long warning before failure occurs. Yet the analysis of Chapter 3 shows that, with modern materials, the proportion of machine failures due to insulation faults are now less than one-third. Furthermore the detection method relies on the most advanced signal processing to extract useful indications, which are then open to interpretation by partial discharge measurement experts. This has made it extremely difficult to increase the confidence of machine operators in the value of this type of monitoring because of their need to refer to differing expert opinion.

Partial discharge monitoring was first applied to isolated insulated components such as bushings and cable stop joints where it has a vital role to play and has been successfully extended to conventional and gas-insulated switchgear. However, when applied to the distributed, multi-path, multi-connection, variably stressed insulation system of an electrical machine winding, as shown in Figures 10.1 and 10.2, it has a much more difficult task and its greatest impact has been on applications such as hydro generators, where a specific failure mode is being searched for in known locations, which allow the precise location of couplers and the tailoring of the signal processing to that failure mode. Work still continues to develop this method, including the use of artificial intelligence (see Chapter 11) to determine the overall deterioration of a winding insulation system but this objective has not yet been reached.

10.7   References

1.   Wilson A., Nye A.E.T. and Hopgood D. On line detection of partial discharges in HV plant. Presented at the 4th BEAMA International Electrical Insulation Conference, Brighton 1982.
2.   Tavner P.J. and Jackson R.J. Coupling of discharge currents between conductors of electrical machines owing to laminated steel core. *IEE Proceedings Part B, Electric Power Applications* 1988; **135**: 295–307.
3.   Harrold R.T., Emery F.T., Murphy F.J. and Drinkut S.A. Radio frequency sensing of incipient arcing faults within large turbine generators. *IEEE Transactions on Power Apparatus and Systems* 1979; **PAS-98**: 1167–73.

4. Emery F.T., Lenderking B.N. and Couch R.D. Turbine-generator on-line diagnostics using RF monitoring. *IEEE Transactions on Power Apparatus and Systems* 1981; **PAS-100**: 4974–82.

5. Harrold R.T. and Emery F.T. Radio frequency diagnostic monitoring of electrical machines. *IEEE Electrical Insulation Magazine* 1986; **2**: 18–24.

6. Timperley J.E. Incipient fault detection through neutral RF monitoring of large rotating machines. *IEEE Transactions on Power Apparatus and Systems* 1983; **PAS-102**: 693–8.

7. Ward D.A. and Exon J.L.T. Using Rogowski coils for transient current measurements. *Engineering Science and Education Journal* 1993; **2**: 105–13.

8. Geary R., Kemp I.J., Wilson A. and Wood J.W. Towards improved calibration in the measurement of partial discharges in rotating machinery. IEEE International Conference on Electrical Insulation, Toronto, 1990.

9. Kurtz M. and Stone G.C. In-service partial discharge testing of generator insulation. *IEEE Transactions of Electrical Insulation* 1979; **EI-14**: 94–100.

10. Kurtz M., Stone G.C., Freeman D., Mulhall V.R. and Lonseth P. Diagnostic testing of generator insulation without service interruption. CIGRE, Paris, 1980.

11. Malik A.K., Cook R.F. and Tavner P.J. The detection of discharges in alternators using wideband radio frequency techniques. Presented at the IEE International Conference on Electrical Machines Design & Applications, London, 1985.

12. Michiguchi Y., Tonisaka S., Izumi S., Watanabe T. and Miyashita I. Development of a collector ring monitor for sparking detection on generators. *IEEE Transactions of Power Apparatus and Systems* 1983; **PAS-102**: 928–33.

13. Stone G.C., Boulter E.A., Culbert I. and Dhirani H. *Electrical Insulation for Rotating Machines, Design, Evaluation, Aging, Testing, and Repair*. New York: Wiley–IEEE Press, 2004.

14. Stone G.C., Sedding H.G., Lloyd B.A. and Gupta B.K. The ability of diagnostic tests to estimate the remaining life of stator insulation. *IEEE Transactions on Energy Conversion* 1988; **EC-3**: 833–41.

15. Sedding H.G. and Stone G.C. A discharge locating probe for rotating machines. *IEEE Electrical Insulation Magazine* 1989; **5**: 14–17.

16. Sedding H.G., Campbell S.R., Stone G.C. and Klempner G.S. A new sensor for detecting partial discharges in operating turbine generators. *IEEE Transactions on Energy Conversion* 1991; **EC-6**: 700–6.

17. Stone G.C., Sedding H.G., Fujimoto, N. and Braun J.M. Practical implementation of ultrawideband partial discharge detectors. *IEEE Transactions on Electrical Insulation* 1992; **EI-27**: 70–81.

18. Wood J. W., Sedding H. G., Hogg W.K., Kemp I.J. and Zhu H. Partial discharges in HV machines; initial considerations for a PD specification. *IEE Proceedings, Part A, Science, Measurement and Technology* 1993; **140**: 409–16.

19. Campbell S.R., Stone G.C., Sedding H.G., Klempner G.S., McDermid W. and Bussey R.G. Practical on-line partial discharge tests for turbine generators and motors. *IEEE Transactions on Energy Conversion* 1994; **EC-9**: 281–7.

20. Stone G.C. and Sedding H.G. In-service evaluation of motor and generator stator windings using partial discharge tests. *IEEE Transactions on Industry Applications* 1995; **IA-31**: 299–303.

21. Kemp I.J., Zhu H., Sedding H.G., Wood J.W. and Hogg W.K. Towards a new partial discharge calibration strategy based on the transfer function of machine stator windings. *IEE Proceedings, Part A, Science, Measurement and Technology* 1996; **143**: 57–62.

22. Stone G.C., Boulter E.A., Culbert I. and Dhirani H. *Electrical Insulation for Rotating Machines, Design, Evaluation, Aging, Testing, and Repair*. New York: Wiley–IEEE Press; 2004.

23. Lyles J.F. and Goodeve T.E. Using diagnostic technology for identifying generator winding needs. *Hydro Review* 1993; June: 58–67.

24. Stone G.C., Sedding H.G. and Costello M. Application of partial discharge testing to motor and generator stator winding maintenance. *IEEE Transactions on Industry Applications* 1996; **IA-32**: 459–64.

25. IEEE Standard. Trial use guide to the measurements of partial discharges in rotating machines. IEEE 1434. IEEE, 2000.

26. Tetrault S.M., Stone G.C. and Stedding H.G. Monitoring partial discharges on 4-kV motor windings. *IEEE Transactions on Industry Applications* 1999; **IA-35**: 682–8.

27. Kheirmand A., Leijon M. and Gubanski S.M. Advances in online monitoring and localization of partial discharges in large rotating machines. *IEEE Transactions on Energy Conversion* 2004; **EC-19**: 53–9.

28. Lee S.B., Younsi K. and Kilman G.B. An online technique for monitoring the insulation condition of AC machine stator windings. *IEEE Transactions on Energy Conversion* 2005; **EC-20**: 737–45.

Chapter 11
Application of artificial intelligence techniques

11.1 Introduction

In earlier chapters of the book we have presented various techniques for condition monitoring. It was seen that most faults do have predictable symptoms during their development, from root cause to failure mode, which can be detected by mechanical, electromagnetic, optical or chemical sensors.

Condition monitoring has to establish a map between input signals and output indications of the machine condition. Classifying machine condition and determining the severity of faults from the input signals have never been easy tasks and they are affected by many factors.

Return to our simple analogy in Chapter 4 of an engineer collecting data and acting upon it. It is in this final, diagnostic stage that experienced engineers still outperform most computerised condition-monitoring systems. Experience and intelligence are extremely important in this interpretative stage when information from different sensors is sifted and condition precisely indicated by tracing the probabilities of different root causes. In recent years, there has been considerable effort to develop artificial intelligence systems that can play the roles currently performed by humans. This is important for at least two reasons.

1. Electrical machines, both motors and generators, are increasingly used as elements in bigger systems where operators are not necessarily experienced in their design.
2. A human expert can be subject to influences that make quick and consistent judgement impossible, particularly when there are many machines in the plant. Correct judgement may also depend on the knowledge and the experience of many experts who are not all available at the same time.

In order to design a computerised system that mimics human intelligence, we should be clear about the response that we expect from experts. They must have a certain amount of knowledge about the relationship between the machine condition and the symptoms. This is the knowledge developed by researchers and engineers, which is what we have been trying to document in this book for rotating electrical machines.

We further expect the experts to apply such knowledge by performing heuristic reasoning upon receiving condition data from the machine being monitored. Development of expert systems in artificial intelligence allows such heuristic reasoning, a feature of human intelligence, to be reproduced in a computerised system. The experts must be able to handle uncertain information and find ways to increase the confidence in their final judgement. Crucial information may come in linguistic format rather than being quantified; for example, 'the water coolant flow of this machine is decreasing much faster than before'. Methods based on fuzzy logic can be used to deal with such situations. Expert systems including fuzzy logic have been used by both machine manufacturers and utilities for condition monitoring, as we will describe in this chapter (see e.g. References 1 and 2).

Manufacturers are adding more functionality to new machines, while the utilities are more focused on aged, existing machines to extend their usable lifetime. It has been realised that human experts facing a condition-monitoring task will receive a large amount of data, some of which may be trivial while others important. The human manner of processing such data features parallelism. Artificial neural networks (ANN), which try to emulate that feature of human intelligence and the learning capability, are particularly attractive in reproducing this parallelism in condition-monitoring tasks. This is being actively researched and we will outline some basic concepts.

Condition monitoring based on fuzzy logic and ANN expert systems can be combined with other methods such as simulated annealing and genetic algorithms, which also emulate human intelligence. A genetic algorithm, for example, does not of itself perform the mapping from input data to an output of machine condition. Instead, it is a stochastic optimisation technique that can be used in the training process of an ANN. Combinations of the techniques and such numerical algorithms will not be described in this book. Interested readers can refer to Filippetti *et al.* [3] for further information. We must point out that it is increasingly recognised that most individual artificial intelligence techniques suffer from specific drawbacks, and that a hybrid approach is needed in most practical situations.

11.2   Expert systems

An expert system treats the condition monitoring of an electric machine like a medical doctor diagnosing a patient. Some initial and superficial information is used to form hypotheses about the machine condition, which are then tested to identify possible faults. Then further evidence is searched for to refine the diagnosis, increase confidence and then make a specific diagnosis. When engineers do this they apply their knowledge, which must be presented to an expert system in a way that can be interpreted by a computer or digital signal processing system. This is not an easy task because such knowledge is usually heuristic in nature, based on 'rule of thumb' rather than absolute certainties. Furthermore human knowledge consists of complex linguistic structures rather than numerical precision, and most of it is almost embedded at a subconscious level. There is no doubt that we need to obtain a substantial amount of knowledge in order to build an expert system that provides quality monitoring functionality. Close interaction between machine and software engineers is necessary, and in order for the final product to be useful, the end user's requirement must also be taken into account.

Knowledge can be presented in a number of ways to the computer or digital signal processing system.

- Frame-based systems represent knowledge in data structures called frames that describe the typical situations or instances of entities.
- Semantic networks represent the knowledge in a graphical structure where nodes represent concepts and beliefs about the system while arcs represent relationships between these concepts and beliefs.
- Rule-based systems represent the knowledge as 'if–then' rules in the computer memory [4].

Typically the rules will not provide certainty in conclusions; rather there will be degrees of certainty in them that will hold for given input conditions. Statistical methods are used to determine these degrees of certainties [5]. Many different representations have been employed in developing expert systems for condition monitoring but the rule-based knowledge scheme is probably the most widely used in practice. Figure 11.1 shows the general architecture of a rule-based expert system.

There is declarative knowledge and imperative knowledge, either of which can be represented as rules.

- Declarative knowledge tells the expert system what it could believe, with a degree of certainty. For example, a negative sequence component above a certain level in the input current of an induction machine implies a stator winding shorted turn fault, or an unbalanced supply.
- On the other hand, imperative knowledge tells the expert system what to do next; for example, looking into other data that may indicate that supply imbalance is not in fact present.

This gives the impression of an ability to apply heuristic reasoning. Human experts are distinguished not only by the quantity of knowledge that they possess but also by the way that the knowledge is applied skilfully. For reasons described at the beginning of this section, an expert system for condition monitoring needs to perform both forward and backward reasoning.

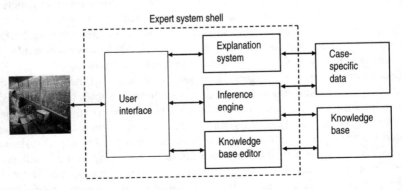

Figure 11.1   Expert system architecture

A feature of the architecture shown in Figure 11.1 is that case-specific data and the general knowledge base are separated. This allows the expert system to be used in different cases provided that the rules in the knowledge base are adequate. They are both separated from a more general part of the architecture called expert system shell. Commercial software shells with the appropriate characteristics to develop the expert system by programming the rules into it and setting the user interface are now available. The shell shown here has an explanation subsystem that allows the expert system to explain its reasoning to the user. The knowledge base editor allows the rules to be modified when necessary and hence makes the expert system so developed maintainable and adaptive. It should be noted that although commercial shells are available, it is the expert knowledge that will determine the quality of the expert system.

Stone and Kapler [1] describe an expert system developed for monitoring stator insulation in turbine generators. Stator windings rated at 4 kV or above experience gradual deterioration of electrical insulation. We have described in Chapter 2 the dominant mechanisms leading to stator winding insulation failure. It can be seen that several interacting factors, including vibration, overheating and moisture contamination, are involved in the degradation process and insulation failure can occur in different parts of the winding. As such factors cause voids in the insulation structure, partial discharge activity occurs leading to further insulation damage. Partial discharge activity can be monitored from the generator or neutral terminals as described in Chapter 10. The main measurement techniques have been described in Sections 10.3–10.5. Partial discharge signals are generally high-frequency spikes, in the 100 MHz range, which can be easily confused with noise from adjacent machines, or arcing or corona activities in the switchyard and switchgear. Expert knowledge is important to ensure the reliability of the partial discharge measurement data, by checking for example the dv/dt rate. Stephan and Laird [2] also confirm this by showing that the measurement data can be contaminated by loose contact of the shaft grounding brushgear. They also show that the data reliability can be improved by considering the correlation of the partial discharge event with the phase angle of the generator terminal voltage. Given the reliability of the acquired partial discharge data, conclusions can only be drawn based on trend analysis, comparison with other generators, maintenance records, earlier off-line test results, present operating point, vibration and temperature measurement, visual inspection and fleet experience. Such aspects are best taken into account using human expert knowledge, represented in rules, as it is extremely hard to establish algorithmic solutions to such problems.

Taken from Reference 1, Figure 11.2 shows an example of the output screen of an expert system monitoring stator winding insulation. It shows the overall risk of winding insulation failure as an 'overall condition assessment', expressed as a relative position between good and bad. This can be used, in comparison with other machines, to determine the machine that is most in need of maintenance. The expert system also provides a list of failure mechanisms that might be occurring in the winding. The probabilities of such mechanisms are indicated and these are estimated statistically using expert experience and knowledge. Other auxiliary information is also displayed

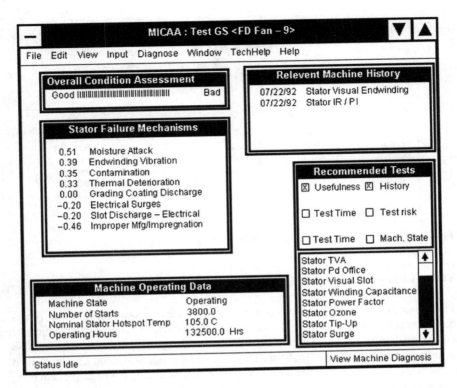

Figure 11.2   Illustration of the output of an expert system used to monitor the integrity of stator winding insulation. [Taken from Stone and Kapler [1]. © IEEE (1997)]

on screen that further aids the operator. About 20 human experts were consulted in developing this expert system.

In practice, expert systems can be made modular with each module focusing on one aspect of the machine condition. All modules are then integrated to provide monitoring over the entire machine. Figure 11.3 shows the architecture of a system developed for steam turbine synchronous generators to look at stator and rotor winding issues. Interactions between different modules are taken into account and these modules may overlap in their functionalities. For example, the condition of shaft grounding gear can affect partial discharge, as mentioned earlier. High end-winding vibration may indicate looseness in the end-winding support, which among other things can cause cracks in the end-winding corona protection, leading to partial discharge in this region. The inclusion of an end-winding vibration monitoring module provides additional information. Combining information from all the modules in a synchronised manner allows for correlation, and the ability to check plausibility of different diagnoses, as described by Stephan and Laird [2].

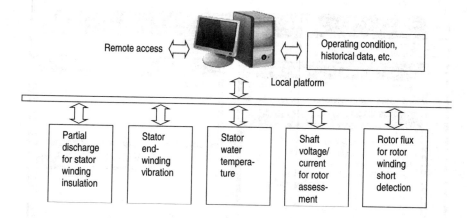

Figure 11.3   Configuration of a modular system

11.3   Fuzzy logic

The range and form of condition-monitoring signatures described in the previous chapters of this book could lead one to the conclusion that interpretation of condition-monitoring signals may be achieved in an unambiguous manner, but this is unfortunately not usually realistic. Taking a stator winding shorted turn fault in an induction motor as an example, an obvious signature would be the negative sequence component in the input current, or acoustic vibration measured on the motor frame. Assuming a balanced voltage supply and perfect manufacture of the machine, such signals can be confidently associated with the stator winding fault as described in Chapter 9. However, supply imbalance and manufacturing imperfections always exist; even a brand new machine would allow the measurement of some negative sequence current and vibration. Unless we can define a threshold value, the information we measure could better be described as 'the negative sequence current is quite large'. The condition of the machine being monitored could also be described in such a linguistic manner, for example the machine condition is not necessarily 'good' or 'bad', but may fall into some intermediate range.

Fuzzy logic is particularly suitable in such circumstances, where the input signals that are to be used for condition monitoring can be associated with certain membership functions. A membership function allows a quantity, such as a negative sequence current, to be associated with a linguistic variable with certain degrees of truth or confidence. For example, Figure 11.4 shows membership functions for the negative sequence current measured on an induction motor that is running in the steady state. The horizontal axis is the negative sequence current normalised to the rated current. The vertical axis indicates how the negative sequence current is associated to four linguistic variables denoted as 'negligible' (N), 'small' (S), 'medium' (M) or 'large' (L). Therefore four membership functions are used. In Boolean logic, the current would be judged either 'true' or 'false' with respect to any of the four linguistic variables. But in fuzzy logic, such a concept is extended to allow the current to be associated with

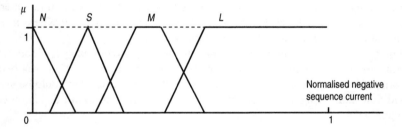

Figure 11.4   Illustration of membership functions for the negative sequence current in a faulted induction motor

these linguistic variables to any degree between 0 (fully false) and 1 (fully true). The same input value can indicate more than one membership function value, implying that different statements can be simultaneously true but to different degrees of truth. For example, a negative sequence current may be judged as 'medium' by a degree of 0.6 while at the same time as 'large' by a degree of 0.3. It is also considered being not 'negligible' or 'small' if the corresponding membership function values are zero. Other input signatures, for example, negative sequence supply voltage, can also be represented in this way, as fuzzy sets by means of membership functions.

Different shapes of membership functions can be chosen for a given application. Construction of the membership functions requires significant insight into the physical meaning of the signals and the linguistic variables to be used. Practical experience and experiments in the laboratory, or the industrial environment is usually the only sure way to gain such insight; the hard work cannot be bypassed, as illustrated by Benbouzid and Nejjari [6]. Nevertheless, there is no doubt that fuzzy logic does provide an alternative and powerful tool of information representation and processing.

Associating input signals with membership functions is called fuzzification. The power of fuzzy logic is more evident in the operation and inference stages of the process, which derive new results that then provide progressively more precise information about the actual condition of the machine. As in Boolean logic, basic fuzzy logic operations could include 'AND', 'OR' and 'NOT', but instead of producing 'true' or 'false' as an outcome, fuzzy logic operations result in a degree value (between 0 and 1) for which the combined statement of the logic operation is true. For example, consider the two statements.

- A: motor negative sequence current is small with membership function of 0.7;
- B: negative sequence supply voltage is negligible with membership function value of 0.5.

One way to define a fuzzy logic 'AND' operation is to use the smaller value of the two involved membership function values. Therefore the degree of the following statement being true is 0.5.

- (A) AND (B): (negative sequence current is small) and (negative sequence voltage is negligible).

Fuzzy logic 'OR' may be similarly defined by using the maximum membership function value, while fuzzy logic 'NOT' may be defined as $1 - u$ where u is the membership function value of the operand involved.

Fuzzy inference is applied to the statements, with or without fuzzy logic operators, to imply the degree to which a consequent statement is supportable. This usually takes the form of 'if–then' rules. For example, one of the many rules that could be used in a condition-monitoring system for stator winding shorted turn fault detection could be expressed as:

Rule (i): If [(negative sequence current is small) and (negative sequence voltage is negligible)] then (shorted turn fault is present)

Because the condition or antecedent of this rule is not 100 per cent true, the output statement, which is also linguistic, is not 100 per cent true either. For the output statement, we are not particularly interested in to what extent it is true. What it means to the actual machine condition is of more importance. We apply a procedure called fuzzy implication for this purpose. This whole process is illustrated in Figure 11.5. In the final plot of Figure 11.5, the horizontal axis is fault severity and the shaded area represents a probabilistic distribution of the fault severity.

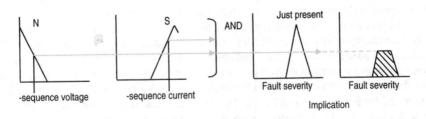

Figure 11.5   *Illustration of the application of fuzzy inference to the negative sequence current as a result of a winding fault on an induction motor*

There may be several rules associated with the same output statement, or consequence, and a condition-monitoring system may have several output statements regarding the machine condition. For instance, another rule output statement could be 'shorted turn fault is serious'.

All rules are then scanned by a fuzzy logic inference engine, generating the corresponding output implications, which are then aggregated to form a single overall fuzzy set regarding the shorted turn fault severity. The order of all the rules being executed does not matter in any way since the aggregation procedure should always be commutative. Three ways of aggregation are commonly applied to the outputs of the rules: maximum selection, probabilistic OR; that is, $a + b - ab$, and summation. Weighting factors can be introduced to reflect the relative importance of each rule. A defuzzification procedure can then be used to obtain a single normalised index for the particular fault that is being monitored. One way is to calculate the centroid of the aggregated overall distribution.

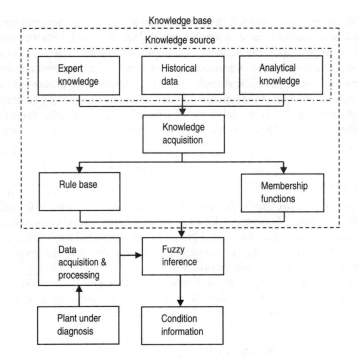

Figure 11.6   Block diagram of a fuzzy diagnostic system

Caldara *et al.* [7] describe an early fuzzy diagnostic system for a linear induction motor drive. The system has the configuration outlined in Figure 11.6. There are similarities between an expert system and a fuzzy diagnostic system in that they are both knowledge-based. The added feature of a fuzzy diagnostic system is in the representation of knowledge as membership functions and the fuzzy inference methodology.

11.4   Artificial neural networks

11.4.1   General

Artificial neural networks (ANN) have been used for many different applications. The feature that is most attractive in condition monitoring is their ability to represent complicated, non-linear relationships and to self-learn in pattern recognition tasks. In monitoring broken rotor bars in cage induction motors for example, a key step is to establish the correlation between the current components and perhaps other signals, and the fault severity. However, a quantitative relationship is complicated and many factors, such as operating point, load characteristic and inherent magnetic or electrical asymmetries in the machine, have effects on the relationship. Such factors are difficult to take into account in any analytical way. A neural network can, however, be trained to represent the desired target relationship. In a neural network approach, the electrical machine is treated essentially as a black box with only the input and output signals used to train the network. The input and output signals are chosen to be those most relevant to the condition-monitoring task, which often requires expert knowledge of the machine. Time domain signals are usually pre-processed by a fast Fourier transform (FFT) and the inputs to the neural network are the FFT components and their derivatives [8]. A neural network properly trained can be used for both fault and fault severity classification, which is the ultimate objective of condition monitoring.

11.4.2   Supervised learning

Of all neural network structures, the multilayered perceptron, trained using the back-propagation algorithm, is perhaps the most widely used. Multilayered perceptron (MLP) training can also be viewed as a gradient descent technique and therefore, to analysts, it possesses a high degree of credibility although the local minimum problem needs to be dealt with in the training process [9].

Figure 11.7 shows an MLP network that was developed by Filippetti *et al.* [10] to detect broken rotor bar faults in induction motors. The network consists of an input layer that accepts six input signals including:

- slip s;
- slip as a ratio of the rated slip s/s_r;
- a quantity dependent on the current component, I_{comp}, at frequency $(1-2\,s)f_{se}$, NI_{comp}/N_mI where N is the number of rotor bars and is normalised to the fundamental current, I, and the maximum bar number N_m, say $N_m = 100$;

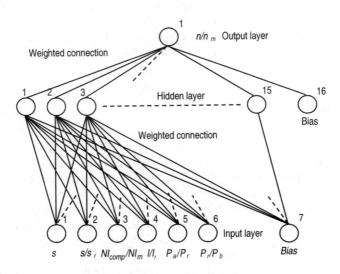

Figure 11.7   A neural network structure for broken rotor bar detection

- f_{se} is the motor supply frequency;
- fundamental current, I, as a ratio to the rated current I_r;
- input power, P_a, as a ratio to the rated power P_r;
- rated power, P_r, as a ratio to a selected base power P_b, say $P_b = 100$ kW.

Above the input layer is a hidden layer of 15 nodes or neurons that each performs the same algebraic operation. Setting the number of hidden neurons is empirical and requires 'trial and error' setting. A common non-linear function of the hidden neurons is the sigmoid 'squashing' function and for a given input to the sigmoid 'squashing' function, x_i, the output of a hidden neuron is calculated as

$$F(x_i) = \frac{1}{1 + e^{-x_i}}, i = 1, 2 \ldots, 15 \tag{11.1}$$

where $x_i = \sum_{j=1}^{7} w_{ij} u_j$ is the weighted summation of all the input signals towards the hidden neuron i, w_{ij} represents the weight from input neuron j to hidden neuron i, and $u_j, j = 1, \ldots, 7$ represents the inputs.

The neural network also has an output layer, consisting of only one neuron in this case, which calculates an estimation of the number of broken bars, n, in a ratio to a maximum number of broken bars, set arbitrarily to $n_m = 10$, for example. The calculation is based on the same function as (11.1) and there are also weighting gains from the hidden neurons to the output neuron. In Figure 11.7, there are two bias nodes that affect the hidden and output neurons directly. The bias nodes can be manually set for calibration purposes.

In order for the neural network structured in this way to have the intelligence that we wish, it must go through a learning process. This enables it to find proper values for the weights in the connections. This requires training data from practical recording, experiments or detailed simulation models that can represent machine defects to a relatively high degree of accuracy. The learning process of a multilayered perceptron network, such as the one shown here, usually requires a form of supervision in which all data are labelled. The data set should cover both the normal conditions and the fault conditions that the neural network system intends to monitor. The desired output for a given input pattern should be explicitly stated in the learning process. The weights are adjusted until the overall error, which measures the difference between the desired outputs and the obtained outputs, is minimised. A typical error function is the sum-of-squares error, as shown below, where N_{data} is the total number of input–output pairs in the data set to be learnt.

$$\text{Error} = \sum_{i=1}^{N_{data}} \{\text{Output}_{calculated} - \text{Output}_{desired}\}^2 \tag{11.2}$$

A way of adjusting the weights is to use the back-propagation training algorithm detailed by many authors including Halpin and Burch [11]. In this algorithm, the error in the final output of the neural network is used to modify the weights, based on a sensitivity-like principle. How quickly the weights can be modified is controlled by two parameters: the learning rate and the momentum that helps to avoid any local minimum and assists convergence. The weights closest to the output layer are modified first and the error is gradually propagated back towards to the input layer to modify other weights. The procedure is iterated until the overall error defined in (11.2) is minimised.

Table 11.1 shows the trained weights that were used in the neural network shown in Figure 11.7 [10]. For instance, the third column is the weights from the third input to the 15 hidden neurons. This input is proportional to the side-band current component at frequency $(1-2\,s)f_{se}$. The corresponding weights have the highest values showing that this input is indeed an important signature for monitoring broken rotor bar faults. Other inputs were used in this network to account for the effects of the operating point of the motor and other inherent asymmetries.

The output of the neural network is proportional to the number of broken rotor bars, as shown in Table 11.2, in a series of cases that were tested using the trained neural network. The actual output value is affected by the setting of $n_m = 10$ in this particular case and the desired output used in the training process. For example, the desired output with one broken rotor bar is 0.1 corresponding to $1/10 = 0.1$. The neural network output should ideally be the same disregarding the operating point of the motor. Table 11.2 shows that in the tests carried out, the obtained output of the trained multilayered perceptron network is very close to the targeted values and the effect of changes to the operating point, with different slips, has been compensated.

11.4.3   Unsupervised learning

A neural network can also learn unsupervised [9]. Compared with supervised learning, an unsupervised system provides a significantly different neural network approach to the identification of abnormal operation and the learning can be more protracted. The system can proceed through the learning stage without provision of correct classifications for each input set of data whereas the connection weight adaptation in the multilayered perceptron is driven precisely by just such knowledge. The possibility of training a network on a set of data, with no labelled inputs or only a fraction of which are labelled clearly, offers significant advantages. However to interpret the outputs so formed and to use them for classification, knowledge of the training data set is still required. Nevertheless, unsupervised networks require fewer iterations in training, since they do not require exact optimisation for their decision-making.

The unsupervised network implementation described here is based on the well-known Kohonen feature map. This feature map approach, presented by Kohonen in the 1980s, derives from the observation that the fine structure of the cerebral cortex in the human brain may be created during learning by algorithms that promote self-organisation. Kohonen's algorithm mimics such processes. It creates a vector quantiser by adjusting the weights from common input nodes to the output nodes, which are arranged in a two-dimensional grid-like network, as shown in Figure 11.8. Such a mapping is usually from a higher-dimensional input space into a low-dimensional output space.

Table 11.1 *Weight matrix for a trained multi-level perceptron network [10]*

| Hidden layer neurons | Input to hidden connection Input layer neurons | | | | | | | Hidden layer neurons | Hidden to output connection Output neuron (n/n_{max}) |
	1(s)	2(s/s_r)	3(N_{lcomp}/N_mI)	4(I/I_r)	5(P_a/P_r)	6(P_r/P_b)	7 (bias)		
1	0.129764	0.28964	1.15775	−0.40374	−0.75320	−0.02838	−0.348574	1	1.00545
2	−0.02786	0.30341	1.03738	−0.36965	−0.72451	−0.18457	−0.27801	2	0.97695
3	1.74813	2.50970	−1.87589	2.51257	4.86890	1.47383	−0.17406	3	−3.99733
4	0.99559	2.99486	5.98608	1.95823	1.19316	0.75494	0.37961	4	4.47211
5	−1.62955	−2.80110	1.83440	−2.33895	−3.92445	−1.828498	−0.11627	5	3.44327
6	0.43743	0.39590	−1.39143	−0.47084	0.26666	0.26027	−0.60810	6	−1.39635
7	−0.49268	0.27708	−0.04326	−0.15611	−0.25054	0.21176	−0.42608	7	−0.06189
8	0.35087	−0.26896	−0.47892	−0.44522	0.50462	0.53946	0.21018	8	−0.89739
9	0.34187	0.116279	−0.52417	0.27934	0.28161	0.31263	−0.65641	9	−0.32928
10	0.20322	−0.04357	−0.22008	0.38878	−0.02762	−0.49045	−0.32304	10	−0.03285
11	−0.35480	−0.13310	−1.33146	−0.71986	0.80925	0.19947	−0.23927	11	−1.51690
12	0.00613	0.64995	1.24199	−0.65056	−1.71787	0.05540	−0.39384	12	1.77122
13	0.33766	0.47155	1.57529	0.70208	0.04751	−0.18017	−0.22157	13	1.47977
14	−0.64187	0.68858	−14.79206	0.96879	−1.58881	−0.54323	−1.64841	14	−9.19390
15	−0.02807	−0.09264	−1.25613	−0.30073	0.29854	0.27405	−0.39073	15	−1.26909
								16 (Bias)	−0.85704

Table 11.2   Test of a trained multi-level perceptron network [10]

Slip (%)	Number of broken rotor bars, n	Targeted output, n/n_m	Obtained output, n/n_m
0.1	0	0.1	0.10031
0.4	0	0.1	0.10017
0.6	0	0.1	0.10020
0.4	1	0.2	0.20200
0.6	1	0.2	0.20000
0.4	2	0.3	0.30500

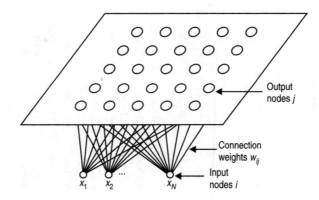

Figure 11.8   Structure of a self-organising feature map

As shown in Figure 11.8, N inputs $X_1, X_2, \ldots, X_N$, are mapped onto M output nodes Y_1, Y_2, Y_M which have been arranged in a matrix; for example, dimensioned 20×20 shown as 5×5 only in figure 11.8. Each input unit i is connected to an output unit j through weight W_{ij}. The continuous valued input patterns and connection weights will give the output node j continuous valued activation value A_j:

$$A_j = \sum_{i=1}^{N} W_{ij}X_i = W_j X \tag{11.3}$$

where $X = [X_1, X_2, \ldots, X_N]$ is the input vector and $W_j = [W_{1j}, W_{2j}, \ldots, W_{Nj}]^T$ is the connection vector for output node j.

A competitive learning rule is used, choosing the winner c as the output node with weight vector closest to the present input pattern vector

$$\|X - W_c\| = \min_j \|X - W_j\| \tag{11.4}$$

where $\| \|$ denotes the Euclidean distance, or norm, between the two vectors. Both the input and connection vectors are normalised.

It is noted that the data presented for the learning purpose consist of only inputs, not outputs. When an input pattern is presented to the neural network without specifying the desired output, the learning process will find a winning output node c according to (11.4), and subsequently updates the weights associated with the winning node and its adjacent nodes. The weights are updated in such a way so that if the same input is presented to the network again immediately, the winning output node will still fall into the same neighbourhood of the output node picked up last time, and with an even greater winning margin when (11.4) is used to select the winning node this time. In other words, the feature of the input pattern is enhanced by updating the corresponding weights. A typical learning rule is

$$\begin{cases} \Delta W_{ij}^{(k)} = \alpha^{(k)} \left[X_i^{(k)} - W_{ij}^{(k)} \right] \\ W_{ij}^{(k+1)} = W_{ij}^{(k)} + \Delta W_{ij}^{(k)} \end{cases} , \quad j \in N_c \qquad (11.5)$$

where superscript k denotes the number of input pattern presented for learning. N_c denotes the neighbourhood around the winning node c with respect to the present input pattern. $\alpha^{(k)}$ is the learning rate that is used in this step of learning.

As more input patterns are presented for the network to learn, the learning rate gradually decreases to guarantee convergence. The initial weights can be randomly set, but normalised, and the input patterns can be iteratively used with several repetitions until convergence is reached [12]. After the training, each output node is associated with a fixed set of weights; that is, $W_j = [W_{1j}, W_{2j}, \ldots, W_{Nj}]^T$. Each input pattern can then be measured for its distance to all the output nodes. If an input pattern is similar to those that contributed to the determination of the weights associated with a particular output node, then the distance between these two input and output nodes will be small. The feature of this input pattern is consequently identified. The structure and the learning scheme of the neural network give rise to its ability of self-organising mapping (SOM).

Penman and Yin [9] describes the details of the construction and learning process of such networks for condition monitoring of induction motors. Figure 11.9 shows the vibration signal feature maps for three machine conditions

- normal,
- unbalanced supply,
- mechanical looseness of frame mounting.

In each case, the input to the neural network is a large number of vibration components obtained from a fast Fourier transform of an accelerometer output. These components are then used to calculate the distances from this input pattern to all of the 20×20 output nodes and the results are shown in Figure 11.9. The ground plane in Figure 11.9 represents the 20×20 output nodes (space), and the vertical axis indicates the distance calculated. It can be seen that the three conditions give distinctively different feature maps. Of course, the physical meanings of such maps depend on the further knowledge that has been embedded in the training process.

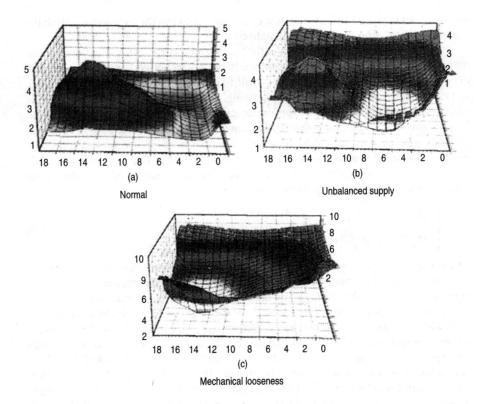

Figure 11.9   Kohonen feature map of machine conditions

11.5   Conclusion

The techniques of artificial intelligence have been seen as potentially valuable for the condition monitoring of electrical machines because the underlying physics of machine operation and dynamics, as shown in Chapters 8 and 9, is so rich in fundamental rules, and these could be exploited by an expert system.

This chapter has shown that these techniques can be used to improve the signal-to-noise ratio in a fault situation, particularly in the complex area of monitoring terminal conditions and the difficult subset of that, the detection and measurement of partial discharge activity. However, the development of artificial intelligence for electrical machine condition monitoring is still in its infancy and despite the considerable work that has been done in this area, much more is required to bring such techniques into the mainstream of condition monitoring.

An abiding lesson from this chapter is to understand the importance in condition monitoring of relating various monitoring signals with one another, what was described in the first edition of this text as multi-parameter monitoring. It is clear, however, that the future of machine condition monitoring will be heavily affected by multi-parameter monitoring and by the application of artificial intelligence to that process.

11.6   References

1. Stone G.C. and Kapler J. Condition based maintenance for the electrical windings of large motor and generators. *Proceedings of the IEEE Pulp and Paper Industry Conference*, Cincinnati, Jun 1997. Piscataway: IEEE; pp. 57–63.
2. Stephan C.E. and Laird T. Condition based maintenance of turbine generators: what makes it real? *Proceedings of the IEEE Electric Machines and Drives Conference*, Madison, Wisconsin, Jun 2003. Piscataway: IEEE. 2: pp. 895–9.
3. Filippetti F., Franceschini G., Tassoni C. and Vas P. Recent developments of induction motor drives fault diagnosis using AI techniques. *IEEE Transactions on Industrial Electronics* 2000; **IE-47**: 994–1004.
4. Birmingham W., Joobbani R. and Kim J. Knowledge-based expert systems and their applications. *Proceedings of the ACM/IEEE 23rd Conference on Design Automation*, Las Vegas, Jun 1986. Piscataway: IEEE. pp. 531–539.
5. Cawsey A. *Essence of Artificial Intelligence*. New York: Prentice-Hall; 1997.
6. Benbouzid M.E.H. and Nejjari H. A simple fuzzy logic approach for induction motors stator condition monitoring. *Proceedings of the IEEE Electric Machines and Drives Conference*, Cambridge, MA, Jun 2001. Piscataway: IEEE. pp. 634–9.
7. Caldara S., Nuccio S., Ricco Galluzzo G. and Trapanese M. A fuzzy diagnostic system: application to linear induction motor drives. *Proceedings of the IEEE Instrumentation and Measurement Technology Conference*, Ottawa, May 1997. Piscataway: IEEE. 1: 257–62.
8. Murray A. and Penman J. Extracting useful higher order features for condition monitoring using artificial neural networks. *IEEE Transactions on Signal Processing* 1997; **SP-45**: 2821–28.
9. Penman J. and Yin C.M. Feasibility of using unsupervised learning, artificial neural networks for the condition monitoring of electrical machines. *IEE Proceedings Part B, Electric Power Applications* 1994; **141**: 317–22.
10. Filippetti F., Franceschini G. and Tassoni C. Neural networks aided on-line diagnostics of induction motor rotor faults. *IEEE Transactions on Industry Applications* 1995; **31**: 892–9.
11. Halpin S.M. and Burch R.F. Applicability of neural networks to industrial and commercial power systems: a tutorial overview. *IEEE Transactions on Industry Applications* 1997; **33**: 1355–61.
12. Wu S. and Chow T.W.S. Induction machine fault detection using SOM-based RBF neural networks. *IEEE Transactions on Industrial Electronics* 2004; **51**: 183–94.

Chapter 12
Condition-based maintenance and asset management

12.1 Introduction

The first objective of condition monitoring is to give early detection of a fault so that avoiding action can be taken, either by shutting the machine down before catastrophic failure occurs, or by taking preventative action that returns the plant to full operational functionality as soon as possible.

So far this book has been concerned with the process of early detection but greater benefits could be achieved from condition monitoring if the information from monitoring is used to schedule maintenance, allowing planned shut-downs so that the life of plant, of which the electrical machine forms a part, can be extended. This is the process of condition-based maintenance, first described in Chapter 1.

However, further benefits could be realised from condition monitoring if the total life-cycle costs of the machine and the plant it serves could be reduced by its application. This in turn requires an estimate of the running costs of plant and forecasts of its variation throughout its life. In the light of this knowledge the plant or asset owner can operate, maintain, renew or dispose of that asset on the basis of the information made available through these processes. This is asset management. This chapter describes how these techniques could be applied to rotating electrical machines.

12.2 Condition-based maintenance

Condition-based maintenance (CBM), is a process of planning plant maintenance action on the basis of condition information obtained from the plant. CBM requires the application of specifically designed software, to be applied to a specific piece of plant, where the plant structure is defined and the software receives condition data from the plant. The CBM software then collates and presents that information in a way that assists maintenance engineers to plan shutdowns. Rao has given a description of CBM applied to electrical machines but this is an enumeration of the condition-monitoring techniques applicable to them [1].

The data which CBM software needs to record when monitoring rotating electrical machines to enable maintenance decisions to be made will be signals of global significance to the machine such as

- running hours,
- number of starts and stops, particularly important for induction motors and standby generators,
- accumulated kWh consumed or produced against time,
- winding and coolant temperatures and in particular temperature rise,
- bearing condition at selected intervals via current measurement and/or vibration,
- for high-voltage machines winding discharge level at selected intervals.

Long-term deteriorating faults that could be avoided by condition-based maintenance are the

- consequences of overheating,
- consequences of overspeed,
- consequences of excessive vibration,
- consequences of shock loads or overloads,
- consequences of corrosion.

An example of the improvement that can be made by the use of CBM in extending the preventive maintenance threshold to plant operators is shown in Figure 12.1.

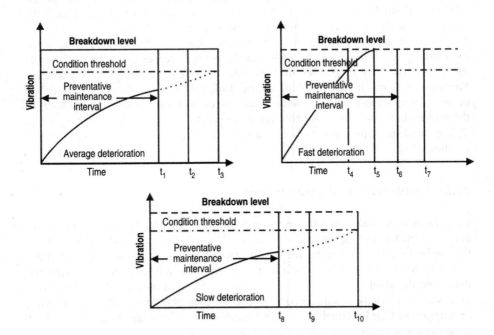

Figure 12.1   The advantage of condition-based maintenance, which can extend the preventive maintenance threshold available to plant operators by expanding the fault-warning period

12.3 Life-cycle costing

Life-cycle costing, also called whole-life costing, is a technique to establish the total cost of ownership, including the capital cost of plant, its installation and monitoring equipment and the cost of maintaining the plant throughout its life. It is a structured management approach that addresses all cost elements and can be used to produce a spend profile on the plant over its anticipated life. The subject was previously hampered by a lack of international standardisation but a standard IEC 60300-3-3:2005 [2] now exists.

The results of a life-cycle costing analysis can be used to assist managers to make decisions where there is a choice of maintenance options; for example, between breakdown, planned or condition-based maintenance. The accuracy of life-cycle costing analysis diminishes as it projects further into the future, so it is most valuable as a comparative tool when the same long-term assumptions are applied to all the options and consequently have the same impact.

Life-cycle costing is an innovation of the last 25 years, reflecting a desire from plant owners to get the maximum value from their investments by sweating their assets. It is born of an operational environment where the costs of capital investment and operations and maintenance labour are high. It is used in developed countries as they attempt to compete with developing countries with lower labour costs. It is also used in privatised national utilities, like railways, electricity, water and gas, as private operators attempt to deliver profitability in a highly regulated environment.

Life-cycle costing raises the incentive of a measurable extension to plant life, reducing overall cost, by investing a proportion of the original plant capital in monitoring and using CBM techniques to reduce through life costs. life-cycle costing has also developed in other industries, for example defence, where high weapon programme costs force governments to demand better field performance, and also aerospace, where the costs of failure due to loss-of-business and litigation can be terminal for private companies.

In order to achieve life-cycle benefits it is necessary to

- understand the life-cycle costs and devise a management system to control them,
- monitor machine condition,
- maintain the machine on the basis of condition where possible,
- assess long-term plant condition on the basis of monitoring signals and manage that plant accordingly.

12.4 Asset management

The final part of the previous section is the process of asset management. This is a business process and forecasting framework for decision-making, based on the behaviour of an asset, which draws from economic and engineering data. Asset management incorporates an economic assessment of trade-offs between alternative investment options, using this information to help make cost-effective investment

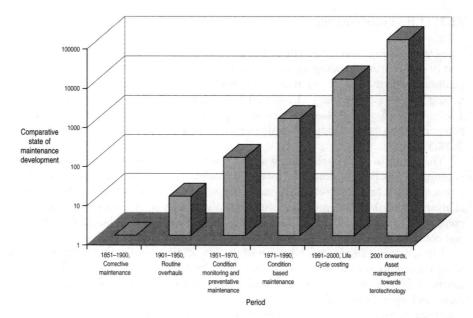

Figure 12.2   Development of maintenance, condition monitoring, condition-based maintenance, life-cycle costing and fully integrated asset management. [Based on Hodges [3]]

decisions and is described by Hodges [3], who gives an interesting time graphic of the evolution of maintenance into what he calls terotechnology (see Figure 12.2).

Proactive asset management exploits condition monitoring to assess the condition of assets, to control their operation, plan maintenance, replacement or disposal. It ensures that maintenance work is carried out only when needed but before assets deteriorate, to avoid costly failures. Asset management has come of age in the last 10 years because of

- changes in the economic environment, as described earlier;
- changes in public expectations;
- advances in instrumentation and computer technology.

Asset management again requires the application of specifically designed software that builds upon that used for CBM, extending it to a network of assets or plant, incorporating life-cycle costing and management information and presents the collated information in a way that assists managers in planning maintenance, replacement or disposal of the assets.

Asset management reflects a level of long-term care for engineering plant that is closer to the health care of human beings by a general practitioner. It is a long way from breakdown maintenance and requires a significant management and resource commitment.

In electrical engineering the greatest advances in asset management have been made with transmission assets with some work going on distribution assets and Brown and Humphrey have given an overview [4] as has Morton [5].

However, the most progress on asset management, in relation to condition monitoring, has been achieved on

- transformers, by the use of transformer gas-in-oil analysis [6],
- transformer winding partial discharge data [7],
- transmission lines by the use of monitoring [8].

One advantage of transformer is that there are no moving parts and the transformer is completely contained within its tank. Therefore condition-monitoring techniques of global significance to the transformer, like gas-in-oil analysis and partial discharge detection, are very effective for asset management and these techniques are particularly applicable to an life-cycle and CBM approach. The asset management of distribution systems has also been considered, as described by Fangxing and Brown [9] and Sonderegger *et al.* [10].

To date, the authors are only aware of the application of asset management techniques to electrical machines in large power stations where key components, such as turbine generator rotors, can be monitored and spares held to be installed in faulty machines when condition-monitoring signals suggest that precautionary measures need to be taken, Kohler *et al.* [11] have given a description of what could be possible.

12.5 Conclusion

The era of condition-based maintenance, life-cycle costing and asset management is relatively new, exemplified by Figure 12.2, and as yet there appear to be no published results to confirm the efficacy of these techniques specifically to rotating electrical machines. However, it is clear from the experience with transmission plant that the techniques can and will be applied to rotating electrical machine, particularly in large installations. Electrical machines are of greatest importance where they are integrated into a large prime mover or drive system, the great benefit of condition-monitoring electrical machine is that by analysing the machine it will be possible to detect deterioration both in the machine and the components attached to it.

Therefore the monitoring techniques that are likely to have the most significance for the assessment of condition for the purposes of maintenance planning are techniques of global significance to the machine such as:

- winding and coolant temperature rise;
- rotor dynamics and bearing condition via power, current, speed and vibration measurement;
- for high-voltage windings partial discharge monitoring.

It is likely that the most effective techniques will in future consider the failure modes and root causes of failure in machines and adopt artificial intelligence techniques, based upon the tables of physical rules exemplified in Chapters 8 and 9, relating various monitoring signals with one another in a multi-parameter approach to give the earliest warning of deteriorating condition. The comprehensive analysis of condition-monitoring signals must take account of the inter-relationship between electrical and mechanical signals.

12.6   References

1. Rao B.K.N. *Handbook of Condition Monitoring*. Oxford, UK: Elsevier, 1996.
2. International Electrotechnical Commission. Dependability management. Part 3-3: Application guide – life cycle costing. IEC 60300-3-3. IEC, 2005.
3. Hodges N.W. *The Economic Management of Physical Assets*. London: Mechanical Engineering Publications; 1996.
4. Brown R.E. and Humphrey B.G. Asset management for transmission and distribution. *IEEE Power and Energy Magazine* 2005; **3**: 39–45.
5. Morton K. Asset management in the electricity supply industry. *Power Engineering Journal* 1999; **13**: 233–40.
6. Dominelli N., Rao A. and Kundur P. Life extension and condition assessment: techniques for an aging utility infrastructure. *IEEE Power and Energy Magazine* 2006; **4**: 24–35.
7. Judd M.D., McArthur S.D.J., McDonald J.R. and Farish O. Intelligent condition monitoring and asset management. Partial discharge monitoring for power transformers. *IEE Power Engineering Journal* 2002; **16**: 297–304.
8. Beehler M.E. Reliability centered maintenance for transmission systems. *IEEE Transactions on Power Delivery* 1997; **PD-12**: 1023–28.
9. Fangxing L. and Brown R.E. A cost-effective approach of prioritizing distribution maintenance based on system reliability. *IEEE Transactions on Power Delivery* 2004; **19**: 439–41.
10. Sonderegger R.C., Henderson D., Bubb S. and Steury J. Distributed asset insight. *IEEE Power and Energy Magazine*, 2004; **2**: 32–9.
11. Kohler J.L., Sottile J. and Trutt F.C. Condition based maintenance of electrical machines. *Proceedings of the IEEE Industry Applications Conference*, 34th IAS Annual Meeting, Phoenix, 1999. Piscataway: IEEE. pp. 205–211.

Appendix

Failure modes and root causes in rotating electrical machines

Based on the machine structure shown in Figures 2.10 and 3.7.

Table 1.

Subassembly	Component	Failure mode	Root cause	Early indicators of the fault
Enclosure	Heat exchanger	Failure of heat exchanger pipework	Defective material Defective installation Corrosion Vibration Shock	Higher winding temperature Higher coolant temperature Moisture and lowered insulation resistance
		Failure of heat exchanger tubes	Defective material Defective installation Corrosion Vibration Shock	Increased winding discharge activity Vibration
	Electrical connections	Insulation failure of connector	Defective material Defective installation Excessive dielectric stress Excessive temperature	Increased connector discharge activity
		Mechanical failure of connector	Defective material Defective installation Vibration Shock	Higher winding temperature Vibration Altered machine performance

Continued

Table 1. Continued

Subassembly	Component	Failure mode	Root cause	Early indicators of the fault
	Bushings	Insulation failure of bushing	Defective material Defective installation Excessive dielectric stress Excessive temperature Vibration Shock	Increased bushing discharge activity
		Mechanical failure of bushing	Defective material Defective installation Vibration Shock	
	Bearings and seals	Loss of lubrication, grease or oil	Lubrication system failure Lack of maintenance Failure of seals	Higher bearing temperature Bearing vibration Bearing noise
		Mechanical failure of bearing element Excessive wear	Loss of lubrication Vibration Shock Overload Loss of lubrication Vibration Shock Overload	
		Electrically provoked failure of bearing element	Shaft voltage Vibration Shock Electrical fault	Higher bearing temperature Bearing vibration Bearing noise Shaft current

Table 2.

Subassembly	Component	Failure mode	Root cause	Early indicators of the fault
Stator winding	Conductors	Failure of conductor cooling hoses	Defective design Defective manufacture Excessive vibration Excessive temperature	Vibration
		Failure of subconductors	Excessive vibration Shock Component failure	Vibration Increased winding discharge activity Increased winding arcing activity
		Failure of conductor bar	Excessive vibration Shock Component failure	Vibration Increased winding discharge activity
	Insulation	Insulation failure of main wall insulation	Defective design Defective manufacture Overtemperature Excessive dielectric stress due to overvoltage Excessive vibration Excessive temperature	Increased winding discharge activity
		Insulation failure of subconductor insulation	Defective design Defective manufacture Overtemperature Excessive dielectric stress, due to high dv/dt Excessive vibration Excessive temperature	

Continued

Table 2. *Continued*

Subassembly	Component	Failure mode	Root cause	Early indicators of the fault
	End winding	End winding insulation failure	Defective design Defective manufacture Excessive temperature Excessive dielectric stress due to overvoltage Excessive dielectric stress, due to high dv/dt Excessive vibration Contamination or debris	Vibration Increased winding discharge activity
		End winding movement or fretting	Defective design Defective manufacture Excessive vibration Contamination or debris	
		End winding contamination	Oil in enclosure Moisture in enclosure Contamination or debris	

Table 3.

Subassembly	Component	Failure mode	Root cause	Early indicators of the fault
Rotor winding	Conductors	Failure of subconductors	Excessive vibration Shock Component failure Excessive temperature	Vibration Increased winding discharge activity Increased winding arcing activity
		Failure of conductor bar	Excessive vibration Shock Component failure Excessive temperature	Vibration Increased winding discharge activity
	Insulation	Insulation failure of main wall insulation	Defective design Defective manufacture Overtemperature Excessive dielectric stress due to overvoltage Excessive vibration Excessive temperature	Increased winding discharge activity
		Insulation failure of subconductor insulation	Defective design Defective manufacture Overtemperature Excessive dielectric stress, due to high dv/dt Excessive vibration Excessive temperature	
	End winding	End winding insulation failure	Defective design Defective manufacture Overtemperature Excessive dielectric stress due to overvoltage Excessive dielectric stress, due to high dv/dt Excessive vibration Excessive temperature Contamination or debris	Vibration Increased winding discharge activity
		End winding movement or fretting	Defective design Defective manufacture Excessive vibration Contamination or debris	

Continued

Table 3. *Continued*

Subassembly	Component	Failure mode	Root cause	Early indicators of the fault
		End winding banding failure	Defective design Defective manufacture Overspeed Overload Excessive vibration Excessive temperature Contamination or debris	

Table 4.

Subassembly	Component	Failure mode	Root cause	Early indicators of the fault
Rotor body	Shaft	Shaft failure	Torsional vibration Shock loading Overload	Cracks located by non-destructive testing or run down tests Vibration
	Rotor core and body	Core hot spot	Defective design Defective manufacture Debris in core Excessive vibration Circulating current Excessive temperature	Higher core temperature
		Core slackening	Defective design Defective manufacture, faulty assembly Excessive vibration Core clamp failure	Vibration
		Integrity failure of body or wedges	Defective design Defective manufacture Excessive vibration	Cracks located by non-destructive testing or run down tests Vibration
	Slip Rings	Sparking Overheating Damaged slip rings	Defective maintenance Defective brushes Excessive temperature	Increased brushgear arcing activity Increased brush wear Increased brushgear temperature
	Commutator	Sparking Overheating Damaged commutator	Defective maintenance Defective brushes Excessive temperature	
	Brushgear	Overheating Damaged brushgear	Defective maintenance Defective brushes Excessive temperature	

Index

accelerometers 92–4, 167–8
AC commutator motors 32
AC reluctance motors 32
ageing of insulation 35–9
aggregate failure rate 68–9
aliasing 113–15
amplifier circuits 102–3
analogue-to-digital conversion 105
artificial intelligence techniques 245–61
    artificial neural networks (ANN) 253–60
    expert systems 246–9
    fuzzy logic 250–3
artificial neural networks (ANN) 253–60
assembly structure 69–71
    distribution of failures 73–5
asset management 265–7
auto-correlation function 116
availability, definition 62
axial leakage flux monitoring 113–17,
    214–17, 223

bearings
    damage caused by shaft voltages 57–8
    failure modes and root causes 270
    failure rates 73, 74
    lubrication oil analysis 152–7
    selection criteria 57
    temperature measurement 128–9
    types 16–18
    vibration response 173–6
bitumen-based insulation 40
brushgear 26
    failure modes and root causes 275
    fault detection 194
    materials 27
busbars 26
bushings
    failure modes and root causes 270
    insulation failures 54–6

cabling, instrumentation 103
cause-and-effect diagram 70
cepstrum analysis 118–20
chemical monitoring 137–57
    detectability 138–42
    gas analysis 148–52
    insulation degradation
        detection 142–52
        mechanisms 137–8
    lubrication oil analysis 152–7
    particulate detection
        chemical analysis 146–8
        core monitors 142–5
circulating current measurement 200
commutators 26
    failure modes and root causes 275
    mechanical frequency components 185
condition-based maintenance (CBM)
    263–4
condition monitoring
    breakdown of tasks 79–81
    definition of terms 61–3
connections: *see* electrical connections
construction of electrical machines 17–33
    connections 26
    enclosures 20–5
    materials used 27
    rotors 18
    stator core and frame 18
    typical sizes and properties 28–9
    windings 18–20
contamination 35
    insulation 38, 43
coolant systems
    blocking 49–50, 56
    chemical monitoring 139–41, 142–52
    heat exchangers 26, 269
    leaks 48–50, 56
    temperature measurement 132, 134
core monitors 142–5

corona discharge 229
correlation analysis 116–17
coupled systems 170
    misalignment 182, 183
cross correlation function 116–17
current densities, typical 29
current imbalance, rotor windings 52,
    211–12
current measurement 99, 100
current monitoring 207–12

data acquisition 104–6
data sampling 105–6
    aliasing 113–15
    discrete Fourier transform 111
    time averaging 120–1
data storage 110
DC generators 32
DC motors
    construction 18, 23
    root causes of faults 32
    rotor winding faults 54
debris measurement
    debris sensitive detectors 100–2
    lubrication oil analysis 152–3, 153–7
definition of terms 61–3
delamination of insulation 38, 40–1
discharge monitoring 229–41
    capacitive coupling method 234–6, 238
    detection problems 238, 239
    earth loop transient method 233
    insulation remanent life determination
        236, 238
    other methods 238
    RF coupling method 231, 238
    wideband RF method 236, 237, 238
discrete Fourier transform 111
displacement transducers 89–91
duration of the failure sequence 62, 63–5

earth leakage fault detection 195–6
earth loop transient monitors 233
electrical ageing, insulation 36–7
electrical connections 26
    failure modes and root causes 269
    faults 54–6
electrical insulation
    contamination 38, 43
    degradation 137–8
        ageing mechanisms 35–9

detectability 138–42
    detection methods 142–52
discharge monitoring 229–41
    to determine remanent life 236, 238
failure modes 39–54, 271–2, 273
    bushings 54–6
    expert systems for monitoring 248–9
    rotor windings 50–4
    stator windings 40–50
root causes of faults 45–54, 271–2, 273
rotor windings 18
stator windings 20
thermal properties 14–15
electrical measurements 99–100
    comprehensive methods 212–21
    discharge monitoring 229–41
    mechanical and electrical interaction 221
    related to specific faults 223
electrical noise, precautions 104, 239
electric loading, typical 29
electromagnetic velocity probes 91–2
embedded temperature measurement 127–31
enclosures 20–5, 26
    failure modes and root causes 269–70
    materials 27
end windings 20, 21
    bracing 20
    failure modes and root causes 48, 272,
        273–4
        insulation 42–3
    materials 27
    stress grading 43
    vibration response 167–8
engine-driven generators 30
environmental ageing 38
environmental conditions 34, 35
    insulation contamination 38, 43
expert systems 246–9

failure, definition 61
failure mechanism, definition 61–2
failure mode and effects analysis (FMEA)
    62–3
failure modes
    bearing faults 56
    connection faults 54–6
    definition 61
    electrical insulation 39–58
    probability density function 66–9
    rotor windings 50–4
    shaft voltages 56–8

stator core faults 54
stator winding faults 45–50
stator winding insulation 40–5
summary tables 269–75
types 66
water coolant faults 56
failure rate 66–9
    aggregate 68–9
    definition 62
    life-cycle variation 68–9
    typical 71–3
failure sequence 63–5
fast Fourier transform 111–12
ferromagnetic techniques, lubrication oil
        analysis 153–4
fibre-optic proximetry 91
fibre-optic temperature sensing 87–8, 131,
        134
fibre-optic torque transducers 96, 97
flame ionisation detectors 149–50
fluidic load cells 97
force measurement 94–7
fuzzy logic 250–3

gas analysis 148–52
gas chromatography 146–7
    off-line analysis 149
    oil degradation detection 153
gearboxes, vibration analysis 118–20
generator exciters 27–8

Hall-effect devices
    current measurement 99
    magnetic flux density measurement 97,
        98–9
    voltage measurement 100
harmonics, converter 113–15
hazard function 62
heat exchangers 26, 269
high-order spectral analysis 115–16
hot-spot measurement 132
hydro generators
    partial discharge analyser (PDA) 239–40
    root causes of faults 30
    typical sizes and properties 29

induction generators 31
induction motors
    construction 18, 23, 24, 25, 28
    failure rates 72, 73, 74

fault detection
    airgap search coils 207
    power spectrum analysis 219
    rotor current monitoring 210–12
    stator current monitoring 207–10
root causes of faults 31
rotor faults 50–2, 185, 223
thermal imaging 133
typical sizes and properties 29
variable speed operation 221
vibration monitoring 181, 183, 186
inductive debris detectors 101, 153–4
inductive torque transducers 96
infra-red gas analyser 152
infra-red thermography 87
insulation: *see* electrical insulation
integrated circuit temperature sensors 87
ionising radiation, insulation ageing 38

key phasors 90
Kohonen feature map 256, 258–60

leakage flux monitoring 113–17, 214–17, 223
life-cycle costing 265
life-cycle variation in failure rates 68–9
lubrication oil analysis 152–7

magnetic drain plugs 154, 156
magnetic flux density measurement 97–9,
        196–200
magnetic loading, typical 29
magneto-elastic torque transducers 97
maintenance regimes 75–6, 263–4
material properties 14–16
    electrical steel 18
    stator core and frame 18
materials, types for various components 27
mean time between failure (MTBF) 62, 71–3
mean time to failure (MTTF) 62
mean time to repair (MTTR) 62
mechanical ageing, insulation 37–8
mechanical loading, typical 29
mechanical properties 14
metal foil-type strain gauge 94–5
misalignment, coupled systems 182, 183
moisture absorption by insulation 37, 38, 49
multilayered perceptron (MLP) training
        254–6
multiplexers 104–5

Nyquist frequency 106

oil whirl 175–6
   frequency components 184, 223
operational conditions 33
operational issues 81
optical debris sensors 101–2
optical encoders
   torque measurement 97
   vibration measurement 91
ozone detection 149

partial discharge analyser (PDA), hydro
      generators 239–40
partial discharges 37
   insulation degradation 138
   internal void discharge 40–1
   slot discharge 41–2
   see also discharge monitoring
particulate detection and analysis 142–8
permanent magnet synchronous machines 31
permeance-wave method 165–7
photo-ionisation detectors 150, 152
'picket fence' effect 112–13
piezoelectric accelerometers 92–4
power (electrical)
   spectrum analysis 217–19
      related to specific faults 223
   typical 29
power spectral density 115
power spectrum 118–20
probability density function 66–9
proximity probes 89–91

quartz thermometers 87

recurrent surge oscillograph (RSO) 204–7
redundancy 70–1
reliability, definition 62
reliability analysis 66–9
reliability function 62, 66
resistance temperature detector (RTD) 82
RF monitors 231, 236, 238
Roebel technique 46–7
Rogowski coils 100, 211–12, 234
rolling element bearings
   mechanical frequency components 184
   vibration monitoring 187–9
   vibration response 173–5

root cause analysis (RCA) 62, 65
root causes
   definition 61
   summary tables 269–75
   types 65
rotational speeds, typical 29
rotors
   airgap eccentricity 56, 75
      detection 183, 213
      producing UMP 165–7, 182
   bearing types 16–18
   cooling 18, 29
   coupled systems 170
      misalignment 182, 183
   electrical angular frequency components of
      faults 223
   failure modes and root causes 275
   materials 27
   mechanical design 18
   mechanical frequency components of
      faults 184, 185

shaft
   dynamic displacement of 185, 223
   eccentricity measurement 90
   static misalignment 184, 223
shaft voltages
   bearing and lubricating oil degradation
      56–8, 153
      monitoring 219–21
   temperature measurement 131
   typical diameters 29
   unbalanced mass 168–71
   vibration response 168–73
      torsional oscillation monitoring 183,
         186
      torsional response 171–3
      transverse response 168–71
rotor windings 18
   broken rotor bars 51
      detection 115–16, 122, 216, 220,
         254–5
      mechanical and electrical frequency
         components 185, 223
   current imbalance 52, 211–12
   failure modes 50–3, 273–4
   failure rates 73, 74
   fault detection
      circulating current measurement 200
      earth leakage 195–6
      fault detection 196–204

inter-turn faults 196–204, 204–7
recurrent surge oscillograph (RSO)
204–7
rotor current monitoring 210–12
stator current monitoring 207–10
materials 27
root causes of faults 50–4, 273–4

sampling rate 105–6
seals, failure modes and root causes 270
search coils
discharge monitoring 234
magnetic flux density measurement 97–8
rotor turn-to-turn fault detection 196–200
stator winding fault detection 194
self-organising mapping (SOM) 259
semiconductor temperature transducers 87
sequence of failure 63–5
shaft
dynamic displacement of 185, 223
eccentricity measurement 90
static misalignment 184, 223
shaft flux monitoring 113–17, 214–17, 223
shaft voltages
damage caused to bearings 57–8, 153
monitoring 219–21
Shannon sampling theorem 105–6
shock pulse monitoring 187–9
shock pulse value (SPV) 187
signal conditioning 102–3
signal processing 109–25
correlation analysis 116–17
spectral analysis 110–16
time averaging 120–1
vibration measurement 118–21
sleeve bearings 175–6
sliprings 26
failure modes and root causes 275
materials 27
slot discharge 41–2
smoke detection 142–5
spectral analysis
aliasing 113–15
basic theory 110–11
correlation analysis 116–17
high-order analysis 115–16
'picket fence' effect 112–13
shaft flux monitoring 214–17
vibration measurement 118–21, 179–82
wavelet analysis 121–5

window functions 111–12
standards
partial discharge measurement 240
permissible limits to residual unbalance
170
vibration 177–8
stator core and frame 18, 20
coolant systems 29
core faults 54
materials 27
typical core length 29
vibration 159–67
electromagnetic force wave 164–7,
182–3
monitoring 182–3
natural frequencies 161–4
stator windings 18–20
coolant systems 29, 48–50
current monitoring 193–4, 217, 223
discharge monitoring 233–4, 236
electrical angular frequency components of
faults 223
electrical structure 230, 232
end windings 20, 21
bracing 20
faults 48
insulation failure modes 42–3
stress grading 43
vibration response 167–8
failure modes 271–2
failure rates 73, 74
insulation failure modes 40–5
delamination and voids 40–1
end windings 42–3
expert systems for monitoring
248–9
inter-turn faults 44
repetitive transients 44–5
slot discharge 41–2
transient voltages 44–5
inter-turn fault detection 182, 183
materials 27
mechanical frequency components of
faults 185
root causes of faults 45–50, 271–2
end winding faults 48
winding conductor faults 46–8
winding coolant system faults
48–50
temperature measurement 127–31
strain gauges 94–6

structural design 69–71
subassemblies 70–1
    distribution of failures 73–5
supply voltage unbalance 33, 158–60, 183
surface tracking 37, 54–6
survivor function 62
synergism between ageing stresses 39

telemetry systems 96
temperature measurement 81–8, 127–34
    bulk measurement 132, 134
    coolant systems 132, 134
    embedded 127–31
    fibre-optic sensing 87–8, 131, 134
    hot-spot measurement 132
    stator windings 127–31
    thermal images 132
    vapour pressure method 131
thermal ageing, electrical insulation 36,
    137–8
thermal images 132
thermal properties 14–16
thermal stress 35
thermistors 86–7
thermocouples 83–6
time averaging 120–1
time between failure (TBF) 62
time to failure (TTF) 62
time to repair (TTR) 62
torque measurement 96–7
torsional oscillation monitoring 183, 186
torsional response 171–3
transient voltages 33, 35, 37, 44–5
trend analysis 120–1
turbine generator analyser (TGA) 240
turbine generators
    construction 16–26
    discharge monitoring 236, 239
    failure rates 72
    off-line fault detection 204–7
    on-line fault detection 195–204
    partial discharge detection 233
    rotor winding faults 52–3
    stator windings 230
    typical sizes and properties 29
    vibration monitoring 181–2, 183, 186
    winding insulation failure, expert systems
        249

ultrasonic discharge detection 236
ultra-violet techniques, chemical analysis
    147–8
unbalanced magnetic pull (UMP) 52, 56, 165,
    182
unbalanced rotor mass 168–71

vapour pressure method of temperature
    measurement 129, 131
variable speed motors 221
velocity transducers 91
vibration measurement 88–94, 176–86
    monitoring techniques 176–86
        faults detectable from the stator force
            wave 182–3
        frequency spectrum monitoring
            179–82
        overall level monitoring 177–9
        torsional oscillation monitoring 183,
            186
    related to specific machine faults 184–5
    signal processing 118–21
    types for various applications 94
vibration response 159–76
    bearings 173–6
    end windings 167–8
    rotors 168–73
    stator core 159–67
vibration standards 177–8, 181
voids, insulation 40–1, 43
voltage measurement 100

wavelet analysis 121–5
wear measurement
    debris sensitive detectors 100–2
    lubrication oil analysis 152–3, 153–7
weights, typical 29
windings 18–20
    materials 27
    see also rotor windings; stator windings
window functions 111–12
wind turbine generators
    cause-and-effect diagram 70
    failure rates 72

x-ray fluorescence detection 157